'VIVA'

Powerful grassroots movements in Latin An   anding fundamental social and political change to a   which has seen
revolutionary governments, authoritarian dictatorships and reformist
military administrations. Through their active involvement women are
seen for the first time as integral to the process of democratization. Yet
these women are not a simple unity with shared aims; class and ethnicity
create division.

Viva explores the growing role of women in the formal and informal
politics of the countries of Latin America. Through contemporary case
studies, the contributors examine how gender-politics in the region is
institutionalized in a variety of spheres varying from the state to local
groups. The book focuses in particular on the role of the state in the
construction of gender, questioning whether the emergence of women's
activism and agendas represent a fundamental shift away from the
historical marginalization of women from politics. The centrality of
gender, class and ethnicity in the ideological construction of 'the nation'
is discussed.

Sarah A. Radcliffe is lecturer in Geography at Royal Holloway and
Bedford New College, University of London. **Sallie Westwood** is senior
lecturer in Sociology at the University of Leicester.

D0139577

INTERNATIONAL STUDIES OF
WOMEN AND PLACE
Edited by Janet Momsen, *University of California at Davis*
and Janice Monk, *University of Arizona*

The Routledge series of *International Studies of Women and Place* describes the diversity and complexity of women's experience around the world, working across different geographies to explore the processes which underlie the construction of gender and the life-worlds of women.

Other titles in the series:

DIFFERENT PLACES, DIFFERENT VOICES
Gender and development in Africa, Asia and Latin America
Edited by Janet H. Momsen and Vivian Kinnaird

FULL CIRCLES
Geographies of women over the life course
Edited by Cindi Katz and Janice Monk

'VIVA'

Women and popular protest in Latin America

Edited by
Sarah A. Radcliffe and Sallie Westwood

London and New York

First published 1993
by Routledge
11 New Fetter Lane, London EC4P 4EE

Simultaneously published in the USA and Canada
by Routledge
29 West 35th Street, New York, NY 10001

Reprinted 1995

Typeset in Baskerville by
J&L Composition, Filey, North Yorkshire
Printed and bound in Great Britain by
Mackays of Chatham PLC, Chatham, Kent

British Library Cataloguing in Publication Data
A catalogue record for this book is available
from the British Library

Library of Congress Cataloguing in Publication Data
A catalogue record for this book is available
from the Library of Congress

ISBN 0-415-07312-X (hbk)
ISBN 0-415-07313-8 (pbk)

To those who died
for life

CONTENTS

CONTENTS

ILLUSTRATIONS AND TABLES

CONTRIBUTORS

Catherine M. Boyle lectures in Spanish and Spanish American Studies, King's College, University of London. She writes about theatre in Latin America, women's writing and cultural expressions, and works mainly in the area of cultural studies, especially through the new magazine *Travesía* (Revista para el análisis y la crónica de la cultura latino-americana), of which she is co-founder (along with John Kraniauskas, David Treece and William Rowe), and whose first number appeared in April 1992. She also works on translating Latin American theatre for performance in Britain.

Nikki Craske is a doctoral candidate in the Department of Government at the University of Essex writing her thesis 'Female Political Mobilization in Urban Mexico: A Case Study of Guadalajara'. She is interested in the issues of women's politicization, popular movements, democratic change and the Confederacion Nacional de Organizaciones Populares (now UNE: Ciudadanos en Movimiento). She has done fieldwork in Mexico and Spain and has an article forthcoming – 'Las mujeres y la CNOP: La Federación de Colonias Populares de Jalisco' – to be published in an anthology on women in urban mobilizations by El Colegio de Mexico, edited by Alesandra Massolo.

Leda Maria Vieira Machado is a Brazilian architect. She developed an interest in women's issues during postgraduate work at the Development Planning Unit of University College London, and gained her PhD in the same department with work on the participation of women in the Health Movement of the Jardim Nordeste Area in Sao Paulo, Brazil. She has published in the *Bulletin of Latin American Research* (1988), and has a chapter in *Women, Human Settlements and Housing* by Moser and Peake, and with the Development Planning Unit. She is currently a lecturer at the Architecture and Planning School of the Belas Artes College in São Paulo.

CONTRIBUTORS

Caroline O. N. Moser is a social anthropologist/social planner, currently Senior Social Policy Specialist in the Urban Development Division, World Bank, Washington DC, after lecturing in Social Administration at the London School of Economics. She has widespread experience of work on gender planning and employment, largely based around her work in Guayaquil, Ecuador. Her articles on gender and development have appeared in journals such as *World Development* and she is the co-editor of *Women, Human Settlements and Housing* (1987 co-edited with Linda Peake).

María-Pilar García Guadilla lectures in sociology at the University of Simón Bolívar in Caracas. She has researched widely in Latin America on women's political participation and has been active in environmentalist politics. She is the editor of *Estado, Ambiente y Sociedad Civil en Venezuela* (1991).

Yvonne Corcoran-Nantes lectures in women's studies at The Flinders University of South Australia. She has researched in Brazil and the UK on women's politics and published widely in journals and through book chapters.

Jennifer Schirmer teaches in the Women's Studies department at Wellesley College. She has researched on women and welfare rights in Scandinavia and is the author of *The Limits of Reform: Women, Capital and Welfare* (1982). She was both a Human Rights Research Fellow and a Research Scholar in the Human Rights and International Legal Studies Programs at Harvard Law School (1983).

Sarah A. Radcliffe lectures on geography in Royal Holloway and Bedford New College, University of London. After having done research in Peru on rural migration and peasant households, she is now working on political mobilization, the emergence of female peasant unionists, and on issues of identity, 'race' and nationalism. Her work has been published widely in book chapters, and journals such as *Society and Space, Journal of Latin American Studies, Bulletin of Latin American Research* and *Review of Radical Political Economy*. Her book on female peasant unionists, *Confederations of Gender*, is due to be published shortly by University of Michigan Press.

Sallie Westwood lectures in sociology at the University of Leicester and has previously conducted research in Ghana, India and the UK in the field of ethnic and gender studies. Her work has been published widely in book chapters, journals and she is the author of *All Day Every Day: Factory and Family in the Making of Women's Lives* (1984; 1985), with Parminder Bhachu *Enterprising Women: Gender, Ethnicity and Economic Relations* (1988) and with E. J. Thomas *Radical Agendas? The Politics of Adult Education* (1991).

ACKNOWLEDGEMENTS

Thinking of the people who have made this book possible means thinking of numerous women and men in both Europe and Latin America whose time, energy and commitment went into this manuscript in one way or another. Our biggest debt of gratitude goes to the contributors to the volume, for their patience with our deadlines and our suggestions, and for their wonderful enthusiasm for the project. Caroline Moser was the initiator and co-organizer of a seminar held in the Institute of Latin American Studies (ILAS), London in November 1989, which was the seed-bed for this collection, and so deserves thanks for her energy and ability in getting the show on the road.

Also, the staff at ILAS, London were unfailingly able to provide a room for the editors when we managed to get together during term. Special thanks to Tony Bell for his quiet assistance, and Leslie Bethell. Also in Britain, we are grateful to the secretarial support in the Department of Geography, Royal Holloway and Bedford New College, especially Kathy Roberts and June Brain and to Stephanie Kneller and Kathryn Baddiley at the University of Leicester. The map was drawn by Justin Jacyno with prompt and careful attention: many thanks. Grants from the Nuffield Foundation to Sarah Radcliffe enabled two trips to Latin America, while a grant from the British Academy allowed for the collection of data on Andean women's historical role in political positions. Sallie Westwood received welcome support from the University of Leicester, Research Board.

In Latin America, the range of women's groups which we (collectively and individually) have learnt from is immense. It is impossible to list them all, or to acknowledge fully the ways in which their struggles and concerns have entered into our writing or research. However, in Lima, the Peruvian Women's Centre 'Flora Tristan', particularly their Red Nacional de la Mujer Rural, and the Centre 'Manuela Ramos' deserve mention as they were always encouraging and interested in our work. Cecilia Blondet provided accommodation and friendship in Lima to Sarah Radcliffe, for which she is extremely grateful, while in Cuzco and Puno the work was not possible without the time given by the departmental federations (FDCC, FDCP), and Amandina Quispe.

At a more personal level (not without its political aspects as well of course), Guy Brown's willingness to look after baby Jessica was of great significance, not least when another deadline loomed. Ali Rattansi was, as always, unfailing in his support and enthusiasm. Our warmest thanks to them both.

Sarah A. Radcliffe and Sallie Westwood

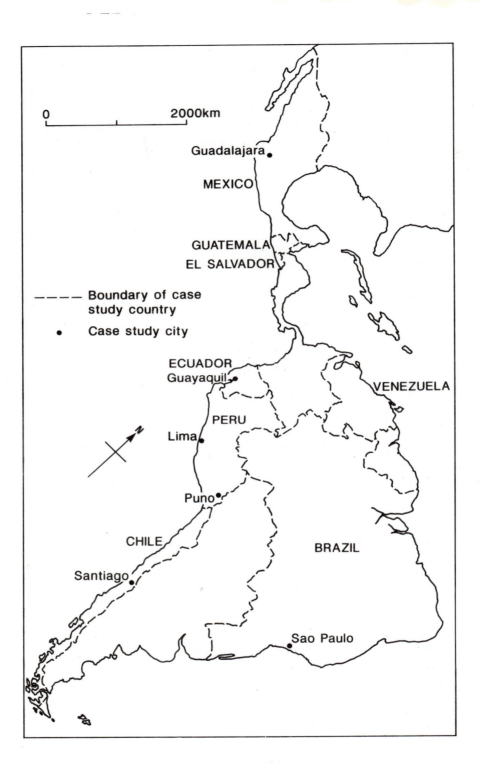

0   2000km

Guadalajara •

MEXICO

GUATEMALA

EL SALVADOR

– – – – Boundary of case
           study country
  •   Case study city

ECUADOR

Guayaquil •

VENEZUELA

PERU

Lima •

Puno •

CHILE

BRAZIL

Santiago •

Sao Paulo •

1

GENDER, RACISM AND THE POLITICS OF IDENTITIES IN LATIN AMERICA

Sallie Westwood and Sarah A. Radcliffe

To think in terms of Latin American women's gender identity also means turning our view to the path of conquest, of colonisation; to how peasant women have been forced into submission; to the slavery of black women; to the historically rooted isolation of middle class women; to the effects of these and other crises in women's lives; to the strong presence of the traditional Catholic church in the lives of many women. In sum, to the traces each and all these experiences leave upon the minds and bodies of this heterogeneous category of women.

(Vargas 1990: 10)

The diversity of gendered identities in Latin America, and the complexity of 'a multidimensionality of situations of power, subordination and exploitation' (ibid.: 9) has been addressed by Latin American scholars and activists, shown clearly in the comment from Virginia Vargas, a leading Peruvian feminist intellectual. Unfortunately, the scope of this agenda for understanding women's movements in Latin America has not been foregrounded by many writers outside that continent (see Eckstein 1986, for example).

While Sonia Alvarez (1990) and others point to the need to place *latina* women's gender and political identities within their specific contexts, this focus has often concentrated on women's 'private/domestic' role, the effects of *marianismo* and *machismo*, and more recently the participation of women in revolutionary formal political structures (Collinson 1990). In other words, research has tended to focus particularly on the 'externalities' of political protest. By contrast in this volume we wish to contextualize women's protest not only in terms of pre-existing political organizations, socio-economic structures and reproductive responsibilities, but also to uncover some of the 'internalities' of political protest, like gender and political identities, images and practices that shape everyday behaviour, symbolism and place in political culture. In other

1

words, we are looking also at (self)representations (de Lauretis 1987) of gendered politics and protest, and how identities are forged within certain 'technologies' (Foucault 1980) of racism and gender (de Lauretis 1987).

In the Latin American context this means examining specific features of economies and societies. These include a European-oriented elite, political and 'high' culture, the Catholic Church, economic policies such as the recent structural adjustment programmes (SAPs) and a highly differentiated popular culture, influenced by the descendants of black Africans who were enslaved, descendants of pre-Colombian peoples, and the mass cultures of the twentieth century (Rowe and Schelling 1991; Roseberry 1989). It also means rejecting the binary opposition between women as self-sacrificing, or as radical guerillas, although women are involved at both these levels. (There is, as yet, only a small literature on guerilla movements and women's participation (see Jaquette 1973; Andreas 1985: 178–87; Labao Reif 1989).)

IMAGINING *AMÉRICAS*

This book will, we hope, introduce a much wider audience to the politics of Latin America while, simultaneously, offering a new theorization of that politics for those already familiar with the region. The region – therein lies our first dilemma. How, and in what ways, is it possible to construct within the imagination the terrain of our discussion? Indeed, the very language with which we conjure the diversity and complexities of this imagined unity – geographically, historically, ethnically and politically – leaves us with a problem because, while we wish to concentrate upon an account which deconstructs Eurocentric views, we are tied, by convention, to many of the terms that promote this unitary vision. Here, in regions labelled Central America, Latin America and more broadly South America, or *América* in Spanish, are a variety of states and nations generated within diverse histories too often suppressed within colonial accounts and perpetuated among the current states largely independent since the 1820s and 1830s. Within these histories are diverse peoples brought together across different times and spaces. This is a huge geographical area comprising mountains, forests, Amazonia, cityscapes and urban sprawl with indigenous and migratory populations including African populations settled by forced migration and enslavement. Historically, however, this 'out-sized reality' (Garcia Marquez quoted in Roseberry 1989: 81) has been homogenized and trivialized by a Eurocentric construction of Latin America or, historically, what was called the 'New World' – it was, of course, new only in relation to a specific perspective from Europe. Thus, William Roseberry, writing on the processes of Americanization comments: 'Part of Latin America's out-sized reality is a multistranded encounter, stretching across nearly five hundred years, with Western powers, the most important of which, since the early twentieth century, has been the United States' (1989: 82).

The colonial encounter, framed by an ideology of 'conquest', was a bloody one marked by genocide, exploitation, brutality and represented within a series of discourses that generated 'the Other' through the creation of an exclusionary language privileging difference that was marked by subordination to Europe in cultural, economic and political terms. The language used to describe the world demonstrates this – the Middle East, the Orient, the New World and the language that was used to socially construct peoples. These accounts have been elegantly excavated in the work of Edward Said (1978) on orientalism and have recently been explored for India by Inden (1990). Similarly, the 'Hispanicist' discourses which produced the 'New World' and the Americas contained within them constructions of 'the Other' which were predicated upon 'fixity' which promoted and sustained stereotypes that remain part of European cultures today. But, as Homi Bhabha (1983: 18) notes: 'Fixity as the sign of cultural/historical/racial difference in the discourse of colonialism, is a paradoxical mode of representation: it connotes rigidity and an unchanging order as well as disorder, degeneracy and daemonic representation.'

The contradictions in colonial discourse, however, did not prevent them from becoming hegemonic and it is with this legacy that our present accounts must engage in order to avoid our own entrapment within these discourses. The current work within subaltern studies (Mani 1984; O'Hanlan 1988, for example) is work of both deconstruction in relation to colonial discourses and recovery – the recovery of the Other, of the knowing colonial subject. The discourses that construct the European ideal and, with this, distance from the ideal are not, however, the exclusive property of the West, for, like capital, they have been globalized and have been appropriated by sections of the Latin American white population. Imperial culture formed cultural configurations in Latin America which generate a play between the processes of 'Otherization' in Europe and its transposition to Latin America, and the outcome of processes of 'Otherization' in the Latin American states. But, increasingly, this does not go unchallenged. The new social movements, generated in the last twenty years, of minority groups, peasants, women and low income groups, have challenged the patterns of emulating Europe which have perpetuated their subordination (Slater 1985; Castells 1978; Jelin 1990).

It is noteworthy that when we first discussed this book in the autumn of 1989 British television was running a series of programmes on Latin America. These programmes represented the current imaginings by the West of the exotica of Latin America: Andean peoples and music,

3

carnival, bands and ceremonial and the abject poverty of large sections of the population (rather than the wealth of the minority). However, even this exoticization, which concentrated upon the link between nation and cultural heritage, reproducing the myth of '*mestizaje*' (or progressive 'whitening' or Europeanization) could not escape the politics of Latin America. Thus, despite the attempts to present Latin America as a land of *coups*, juntas, ceremonial and samba, the presentation of materials on Colombia and the cocaine barons, or the poverty of so many peasants and urban dwellers, consistently raised political questions that are directed as much at the West as at the incumbent elites of the Latin American states. Thus, a more recent series, dubiously entitled 'The "savage" strikes back' provides a platform for the voices of indigenous peoples in the Americas. The series opened with a programme which featured the struggle by indigenous Indians in Ecuador against the power of multinational oil companies. Homogenization within the accounts of the continent is, thereby, within the popular media, giving way to the complexities and diversities of the peoples and nations and states within the region.

The exoticization of aspects of the region is one element in the racism which continues to reproduce representations of the peoples of the continent from within Eurocentric discourses. We are mindful of our responsibilities as white Western women, as women of *el Norte*, in relation to the work of deconstruction and in not simply providing space for Latin American women to be heard, but also shaping the ways in which those spaces will be constructed. Just as the apparently neutral presentation of 'testimonies' particularly favoured in Western presentations of Latin American experience in recent years is, in fact, political (see LAP 1991; Patai 1988) so, too, is the presentation of women's experience of popular protest a highly politicized and selective process. Jennifer Schirmer in Chapter 2 addresses these issues in relation to the CoMadres of El Salvador and the CONAVIGUA widows of Guatemala.

We are equally mindful that all the contributors to the book, whether Brazilian, Venezuelan, North American or British, inhabit not simply national identities but subject specialisms as well, which in themselves multiply the numbers of discourses within which we construct our discussions. Most significantly, these disciplines are the outcome of the post-Enlightenment period of 'modernity'. Thus, geography, sociology and political science have grown up as part of the Modern project based on rationalism and a belief in the relationship between knowledge and progress on a Eurocentric model which is often viewed as the opposite of the 'magical consciousness' of 'the Other'. Thus, we are aware that it is not a simple or innocent matter to open up spaces wherein the voices of the women of the South may be heard – nothing is so simple. Politics inevitably re-enters the terrain of our work rather than it being constructed as an external reality upon which we comment. This, of course, has always been one of the central premises of feminism: it is not surprising, therefore, that feminist discourses should be important to a volume which addresses issues of gender, racism and political identities.

Feminist discourses have, however, been similarly and correctly the subject of scrutiny in relation to their universalizing and essentializing accounts of 'woman'. Black and minority women in the West have raised their voices against a white middle-class account of feminism and have been joined by 'Third World' women. Both groups have denied the relevance of Western feminisms, at the same time as they have sought ways to appropriate some part of its politics. In the Latin American states, feminisms have flourished amid debate across similar constraints of 'race' and class as in the West (Jaquette 1989; Andreas 1985; Butler Flora 1982; Vargas 1990). Leila Gonzales, a Black Latin American feminist, accuses the feminisms in the continent of being 'racist by omission':

> Latin American feminism loses much of its force by making abstract a fact of great importance: the pluricultural and multi-racial character of societies of the region. ... To speak of the oppression of the Latin American woman is to speak of a generality which hides the hard reality lived by millions of black and indigenous women.
>
> (Gonzales 1988: 97)

As white Western women we are mindful of the critiques generated by black and minority women and by women of the South especially in relation to the ways in which they are represented in feminist and social science discourses (see Zabaleta 1986). Chandra Talpade Mohanty (1988) has examined this in relation to a series of texts on South Asian women and the ways in which Western eyes and scholarship reconstruct Indian women's lives in South Asia within discourses that are redolent with those of the colonial era. These colonizing discourses are not simply 'without', in the eyes of the West alone, but also 'within', and among, the dominant classes of the Latin American states and within the discourses produced and reproduced by the state around masculinities and femininities in these countries. It is precisely this complex articulation of 'within' and 'without' that is raised in this volume. Our task has been to deconstruct the homogenizing, universalizing account of Latin America which contributes to the exoticization and 'Otherization' of the peoples of Latin America, and, instead, to give full attention to diversity and the specific ways in which racisms, gender and class relations are articulated in the different states of the Americas.

5

RACISMS AND THE RACIALIZATION OF LANGUAGE

European racisms are both similar in relation to the colonial legacy and distinctive, marked by the complexities that have been part of Europe and including, in relation to Spain and Portugal in particular, the presence of black Africans and Jews and the contestations between Christianity, Judaism and Islam. Spanish and Portuguese imperialism reinforced a sense of the specificities of European identities, however fictitious, and amplified the processes of 'Otherization' in relation to the indigenous populations and those enslaved and transported to South America. As Stuart Hall notes:

> The story of European identity is often told as if it had no exterior. But this tells us more about how cultural identities are constructed – as 'imagined communities', through the marking of difference with others – than it does about the actual relations of unequal exchange and uneven development through which a common European identity was forged.
>
> (Hall 1991: 18)

Such a process has had a profound impact on the cultural development of Latin America and on the ways in which gender identities and relations are shaped by racisms in Latin American societies, whether in forms of hegemonic gendering of the Hispanic *marianismo* figure, or the subaltern masculinities and femininities associated with subordinate groups. Gender is lived through racisms and social constructions of 'race' in Latin America. In Brazil the women's movement is cross-cut by racism as well as by class. Thus, opportunities for women in the modernizing period from the 1960s onwards were opportunities for 'whiter', urban women. By contrast, for most working women concentrated in domestic service, who tend to be of African descent, these opportunities remain remote (Sarti 1989: 77).

Constructing 'the Other' has, in part, been about the ways in which the diversity of colonial peoples has been re-presented in relation to racial categorizations. In the Latin American states, especially Brazil, this has been organized around an ideological configuration known as 'racial democracy' which is, in effect, the process of 'whitening' through miscegenation. Moreover, gender and 'race' were brought together in the crucial figure of the mulatto woman. Peter Wade concludes his discussion of Colombia:

> The emergence of a large mixed intermediate group ... has established the myth of a Latin American 'racial democracy' based on the predominance of the mestizo and the mulatto and in which racial marks are no barriers to marriage and social mobility. It is important to recognize, however, that the mechanisms of racial and

social vertical mobility that exist in Latin American societies draw their dynamic from an attempt to escape blackness that has been and continues to be negatively evaluated, and thus to whiten oneself and eventually the population as a whole.

(Wade 1986: 16)

Racism in Andean countries follows a similar trajectory, whereby those groups with perceived indigenous heritage have been perceived negatively throughout post-contact history (Bourricaud 1975; van den Berghe and Primov 1977). In recent times governments have promoted *mestizaje* within a pro-indigenist populist discourse, in which General Lara of Ecuador could claim, 'everyone is an Ecuadorean now'. Again, the production and reproduction of *mestizaje* as an active mobilizing force is highly gendered with peasant ('indigenous') women being subject to pressures to adopt Western cultures in sex-segregated domestic work (Radcliffe 1990), while men receive parallel treatment in the army and the informal sector. The process of an 'escape' from blackness is marked by violence, a violence which forces people away from their cultural roots.

Abdias do Nascimento has drawn attention to this in relation to the black population of Brazil and has provided a new narrative which gives a central place to black people rather than, as he points out, ignoring the specific identity of those Brazilians of African descent. Attacking the notion of a 'racial democracy', he notes that:

The attempt was to silence millions of Brazilians of African origin with the illusion that, by solving the dichotomy between rich and poor or between worker and employer, all racial problems would be automatically resolved. This position of the white Eurocentric ruling elite was taken to the extreme of elaborating an ideology called 'racial democracy' whose goal was to proclaim the virtues of Brazilian race relations, presenting them as an example to be followed by the rest of the world.

(Nascimento 1989)

Nascimento's account emphasizes the importance of slavery in the development of Brazil and the ways in which black people resisted enslavement and the economic and political brutality of the colonial era while developing their own infrastructure for political struggle. Slavery was not abolished until 1888 but as Nascimento (ibid: 42) comments: 'The enslaved African became a "citizen" as stated under the law, but he also became a "nigger": cornered from all sides.' In Nascimento's revision of Brazilian history, he places the fate of African Brazilians firmly on the agenda. He shows that the migrations from Europe in the early part of the twentieth century displaced the black population, by economically favouring whites over blacks and further reinforcing white racisms and the position of the black populations.

Nascimento's account is unusual in that, despite the continuing impoverished position the black populations still occupy, very little attention is paid in Latin American literature to the issue of racism and ethnicities in Brazilian society and within Latin America more generally. However, this situation is changing (see Winant 1992) especially in relation to the widening coverage of the fate of indigenous populations in Latin America, particularly in Brazil where discussion of Amazonia cannot fail to engage with issues of racism, 'Otherization' and ethnicities.

The fate of the indigenous peoples of South America has caused some indigenous leaders to coin the term 'fourth world' for the space they occupy on the world stage. Indigenous groups have, until recently (according to the West) been 'the forgotten peoples'. But, there is now a move across the Americas to unite the indigenous peoples from the Arctic North to Amazonia. Again, Brazil, with its complex and diverse populations and its ideology of 'racial democracy', provides key indices of the processes of racialization. The Indian population was to become Brazilian via miscegenation, but those who maintained their Indian status, like the tribal Yanomami peoples, remained outside the nation and subject to genocidal practices which have annexed their lands and stolen their livelihoods (MacDonald 1991).

The racialization of the populations of Latin America is demonstrated through language: mestizo/a is the acceptable face of the Latin American populations, a term which is bound to the processes of 'whitening' and which of itself maintains the Eurocentric view of the world. By contrast, 'Indios' is a pejorative term used to denote backwardness and with this marginalization (Stark 1981). It is a familiar discourse and one which is deeply embedded within the fabric of societies, institutionally and in the commonsense world of everyday encounters. The implications of this racialized language for political mobilization are to be found in the ways in which 'ethnic' identities can be used to promote or suppress political identities. This is clearly expressed by Rigoberta Menchú who states (1983: 166) 'I am an IndianIST, not just an Indian. I'm an Indianist to my fingertips and I defend everything to do with my ancestors.'

As the chapters in this book show, especially Sarah Radcliffe's on Peru and Jennifer Schirmer's on Guatemala and El Salvador, ethnic identities and class identities are always constructed and refracted through gender identities. This is a complex politics in which subaltern men and women can be subject to a 'shared powerlessness' which heightens common identifications in the face of Hispanic culture and politics. As a Guatemalan woman leader of the IXQUIC group says: 'It is important for Western women to understand why we don't have the same demands (as them) . . . what could we ask for now? In many senses for

Guatemalan women it would mean equal repression and we already have that' (quoted in ISIS 1986: 23).

The language issue has a very specific resonance within the context of Andean cultures where the complex colonial encounter is represented as that between written and spoken languages: between the Andean word and the Hispanic book symbolized through the story of Atahualpa, the Inka chief, and his encounter with the Dominican Fray Vicente Valverde carrying the bible which fell, or was pushed unceremoniously, to the ground (Seed 1991). Symbolically, it was the encounter between indigenous peoples and the colonizers, and between Quechua and its oral tradition, and Spanish and its text (Howard-Malverde 1990). It is an ongoing struggle that has been organized around this binary opposition.

There is, however, no simple opposition within or between political identities in relation to the ways in which language positions subjects. Indeed, in the struggles between landlords and peasants in Peruvian haciendas over appropriation and control of resources, the issue of labelling peasants was integral to the power relations between the two groups. In a recent account of such a struggle, Fiona Wilson notes the racialization of language and political identities when she argues that:

> Identities were especially fluid and ambiguous in the 1940s a time of social ferment and economic chaos. This fluidity was borne out by the fact that hacendados (estate owners) could never fix on a single term to define the people living on their properties. They were described as socios (members), operarios (operators), braceros (workers/labourers), trabajadores (workers), colonos (renters), arrendatarios (sharecroppers) indigenas (indegenes), indios (Indians), niños (children), and gente (people). . . . The shifting names given to the protagonists symbolized their shifting identities.
> (Wilson 1986: 86)

We are suggesting that language helped to construct the shifting terrain of identities which included the perjorative *indios* and its paternalistic cousin, *niños*. What is noteworthy is that none of the terms dignifies the ethnic origins of the people involved, a situation which has endured to the present day. What Nascimento foregrounds in relation to the African Brazilians is reproduced in modern Peru through the adoption of *campesino/a* (peasant), a political-economic class identity with the ethnic referent removed, made invisible. The term *campesino* is used in this book by convention but with the ethnic specificities reinserted.

The symbolic use of 'peasants' and 'workers' is also part of a public language in relation to the nations of Latin America and their construction in the twentieth century. Nationalist discourses are generally organized around very specific ethnicities which privilege a Eurocentric

9

account of the nation against the indigenous voices that have to fight for a place in the nation. This, in part, reproduces the opposition between the literary and the oral traditions in which the former is privileged in nation-building in Anderson's (1983) account (Rowe and Schelling 1991).

NATIONS AND STATES

To invoke nations is to foreground not the commonly assumed geographical inertia of state boundaries, but the terrain of national identities and the struggles around these which are fought over throughout South America. A symbolic figure such as Simón Bolívar, one of the leaders of South American independence and advocate of a Pan-American movement, retains a contradictory place in mythologies about the nation. Different constituencies interpret and contest the origins and role of Simón Bolívar as 'father' of the nation in Bolivia and Venezuela, for example (Rowe and Schelling 1991). To consider questions of ethnicities, racisms and gender opens up the terrain of nations, nationalisms and national identities (Yuval-Davis and Anthias 1989). Nations, on the other hand, assume closure and exclusion. To speak of nations is to invoke 'imagined communities' (Anderson 1983) that are no less powerful or real for being symbolic (Hall 1987). Gramsci's now familiar analysis of ideological hegemony helps us to understand the ways in which the nation is constructed through a consensus around 'the people'. The ideological construction of the nation is organized around specific ethnicities and constructs 'fictive ethnicities' and national identities as a crucial part of the national story. Thus, in Argentina the Indian presence was both forced from the land and made invisible in terms of the national consciousness (in ways similar to the exclusion of African Brazilians in Brazil) through a series of discourses that wove together the history of Argentina, in such a way that the ethnic diversities of the nation were written out.

In the recent histories of Latin American states, nationalism and nationhood were key rationales for the interventions by the military during the 1970s. These interventions defined the state and its power as the nation and national interest, at the same time as 'the people' were seen as dangerous. Although in Peru the military intervention did have a more progressive character, military interventions generally, especially in the southern Cone, were tied to levels of state repression that remain firmly entrenched in the Western popular imagination as the key indicator of politics in the region. This is, in part, because the symbols of nationhood were militaristic, redolent with masculinity and, therefore, highly gendered. The generals justified seizure of power on the basis of their privileged position to push through 'national development'. The military coexisted with what was characterized as a *comprador bourgeoisie* (a term little used now) in control of national resources including land, tin, coal, but also tied to the world markets and the life-styles of the bourgeoisie in the metropolitan core. This power bloc was often an uneasy alliance between political and economic interests in which the militaristic state engaged in the 'management of populations' through terror, while class interests secured the impoverishment of the mass of the population. However, despite the creation of regimes of terror and of increasing wealth for upper- and middle-class groups, the nation was not secure; fractions, *coups*, splits within the bourgeoisie were one problem, as was organized resistance from diverse sections of the population – workers, peasants, intellectuals, indigenous peoples, the urban poor and women with diverse backgrounds. In the 1970s, the famous strikes in the Bolivian tin mines and the political organization among Chilean workers, for example, prefigured a different trajectory. Citizenship and civil society were in this period repressed by the state and the organs of civil society tried to survive at a subterranean level – they went underground and people went into exile.

A more subtle and discriminating version of authoritarian models of nationalism was to be found in authoritarian populism, the most famous example being Peronism in Argentina, which sought to forge national identities by welding certain sections of the populations into 'the people' who were then identified via the leader Perón with the national interest, national culture and nationhood. Trade unionists (of the unofficial variety), socialists, women's organizations, peasant organizations and ethnic identities were perceived by this national project to be outside the nation, sources of danger because they had an alternative class–gender–ethnic perspective from which to generate their own version of populism and thereby a degree of autonomy from the view on offer.

We have seen this alternative populist project mobilized during the 1980s across the states of Latin America. They could not be interpellated as 'the people' on a definition from above. Complexities continue, however; authoritarian populism uses the symbols of populism to the very populations that have generated them. As Ernesto Laclau (1987: 161) succinctly put it: 'The ideology of the dominant class, precisely because it is dominant, interpellates not only members of that class but also members of the dominated classes.' This is highlighted by the continuing authority granted to *gauchero* figures in Argentinian politics, with the election in 1989 of Carlos Menem, self-conscious imitator of the evocative nationalist symbol Facundo Quiroga.

In this way, authoritarian populism ties a specific class interest (or section of the bourgeoisie) to the national interest; for it to succeed ideologically its class interest is, therefore, submerged in the political discourse, and ethnicity, nation and gender are far more likely to be

11

privileged in the construction of nationalism. Thus, in Latin American history, men and masculinity are tied to the defence of the nation and the protection of family, home and the people, while women are cast not as defenders but as reproducers of the nation as wives and mothers (as occurs elsewhere in the world, see Yuval-Davis and Anthias 1989). Thus, as discussed in Catherine Boyle's chapter, in Chile, middle-class housewives were mobilized against President Allende (1970–3) over food shortages and came on to the streets to bang their saucepans.

As previous work has shown, the state has a major role to play in articulating masculinities, femininities and the role of women through a series of discourses which position women as family members, mothers and as class subjects (Alvarez 1990). But it is also the case that these discourses are racialized and thereby privilege specific ethnicities. In Brazil or Argentina only certain women are celebrated as mothers of the nation and idealized against the position of Indian and African Brazilian women. In addition, the state organizes via the legal system the ethnic and gender boundaries of the nation; in other words, citizenship is defined in relation to certain privileged groups as well as being transferable or inheritable through men rather than women. In this way, although motherhood may be celebrated within the nation and by the state, mothers may not confer equal citizenship on their children.

Equally, the state and the political parties sought, and still seek, to articulate discourses around the family and motherhood in relation to the powerful Roman Catholic Church in Latin America. Eva Perón was the most famous example of this contradiction, a woman with charisma and power who consistently defined herself in relation to motherhood, as mother of the nation, and as a wife and helper to Perón. As the chapters in this volume demonstrate, most especially that by Leda Machado, since the Medellín declaration in Colombia, Catholicism, in its 'option for the poor', has proved very contradictory – a force as much for organizing opposition to the state as for reproducing the status quo. The symbol of the mother in the Latin American states has been used both to signal opposition and resistance to the excesses of the state, shown clearly in Jennifer Schirmer's chapter, and to represent the epitome of the national subject. Housewives have organized in House-wives Committees to support progressive resistance, like the Bolivian Tin Miners Against the State (Domitila Barrios de Chungara 1978). This means that the state does not have familial ideologies all its own way – the family is deeply and consistently contested terrain in Latin America and the ways in which it is used by the state meet with popular resistance (explored in Sarah Radcliffe's chapter on Peru and Catherine Boyle's in relation to representations in Chile).

The authoritarian populism of Peronism in Argentina and Vargas in Brazil gave way to new settlements in the 1970s and 1980s that

Guillermo O'Donnell has characterized as Bureaucratic Authoritarian (BA) regimes. In industrially advanced countries such as Brazil, Argentina, Chile and Uruguay the military had a new role in the development process. To overcome a perceived crisis, a new model for growth was instituted that required a high degree of repression via the military as well as the involvement of technocrats and state functionaries within the state bureaucracies. However, the economic model of BA fails to take sufficient account of the legitimation crises suffered by the military regimes in these states. It is also important to separate out the case of Peru where, during the 1970s, a far more progressive and populist military government was in evidence. Other writers have highlighted how BA regimes shaped the dynamics and concerns of feminisms throughout Latin America (see Jaquette 1989).

The state has been, and is, a very powerful force in the politics of Latin America and it is very easy to slip into an account of the state that is undertheorized and that unwittingly constructs the state as a simple unity, or a series of institutions with functions and thereby some form of agency. The state, on this model, becomes a political actor with interests and there is some suggestion of this in Sonia Alvarez's rather functionalist account of the state in Brazil when she writes: 'Thus, one of the central functions of the State is to promote changes in the one system of exploitation (patriarchy or capitalism) to suit the needs of the other' (Alvarez 1990: 30). Such a view also reproduces a familiar account of the state as referee in order to secure the processes of reproduction under capitalism, and thereby to maintain the status quo.

In contrast, our account would wish to suggest a disaggregation of the state as a way of exploring the contestations within and around the state, and thereby suggesting, even in bureaucratic authoritarian regimes, that 'the state' is contested terrain in which power blocs seek to impose an agenda of development and control in the social formation. Clearly, the consequences of these struggles are highly variable with, for example, a higher degree of consistency within the Brazilian state than the Argentinian or Peruvian during the 1970s (see Slater 1985). The state articulates the exclusivity of the nation via the legal requirements surrounding citizenship and thereby it interpellates 'citizens', but may also find those not included contest their exclusion or those subjected to state abuses answer the state from within the state's own rhetoric about the protection of nationals and the rights of citizens. This is precisely what women across Latin America did in relation to the repression of the 1970s and early 1980s. (For case studies of this period see Schmink 1986; Navarro 1989; Chinchilla 1979; Viezzer 1979; Cahiers des Amériques Latines 1982; Chaney 1979; Mcgee 1981 and Sarti 1989.)

EMBODIMENTS OF POLITICS AND IDENTITIES

Central to many of the accounts in this volume is the issue of the human body and the ways in which it becomes, in a Foucauldian sense, the site of technologies of domination and modes of resistance. Whether in terms of gender, racisms or class, it is the body which acts as signifier of difference and through which individuals identify themselves and others in the constitution of their politics. Especially in Latin America it is still the case that the 'naturalization' of difference is tied to the body. The signifier of colour is a crucial aspect of the ways in which racial formations are generated and sustained, for example in Brazil, but also more generally in relation to the ideology of racial democracy which has at its core, 'whitening'. As William Rowe and Vivien Schelling comment:

> *Mestizaje*, a word denoting racial mixture, assumes a synthesis of cultures where none is eradicated. The difficulty with the idea of mestizajes is that, without an analysis of power structures, it becomes an ideology of racial harmony which obscures the actual holding of power by a particular group.
>
> (Rowe and Schelling 1991: 18)

Not only is 'race' tied very specifically to the body, but these bodies are at the same time, always, already gendered and class-placed. There is, therefore, a very important sense in which popular politics have been tied to the ideological and social construction of the nation around questions of the body and the 'management of populations'. Low-income groups, subordinated 'ethnic' groups and diverse women have all had to struggle against policies controlling their fertility, health and freedom of movement of the body. Violence under military regimes has been directed against individuals who epitomized the 'other', the threat to the nation. Thus, in countries with emergency law such as Peru and Guatemala at the present time, peasant women are subject to rape and torture by national soldiers in their actions against 'terrorists' (Butler Flora 1982).

In response to these 'embodying' violences, popular cultures have generated mythical figures in which, in a series of reversals, white, Hispanic males become the incarnation of evil. For example, in Lima in November 1988, white, blond men driving big cars were said to be taking out children's eyes in order to sell them. The *sacaojos* ('eye pluckers') expressed popular fears that children of low-income populations could look out on no future in the aftermath of harsh austerity measures (September 1988) and rising political violence (Vargas 1990: 36–7).

At the level of dominant discourse however, women have been placed within nationalist imagery as 'mothers of the nation', symbolically and

14

biologically. As discussed above, it is clear from the 'national stories' that not all women are equally 'mothers' of the nation. On the contrary, on the very terrain of biology which essentialist feminisms would claim for women, racisms separate women and their experiences. For example, it is black women in Brazil who have been the subjects of forced sterilization. Struggles over fertility, the body and group cohesiveness are initiated in response to both right-wing and left-wing political groups. Thus, in Central American countries and Brazil, women of diverse cultural and ethnic groups face extinction by death squads and introduced diseases. In Peru, by contrast, the radical leftist group Shining Path accuses feminists of depriving the revolution of support by encouraging 'popular' women to limit child-bearing (ibid.: 35).

In other ways, too, the state has engaged, through its personnel, not simply in the management of populations, but also in the annihilation of certain sections of its subject groups and the torture and abuse of men, women and children. Under military regimes in the 1970s and early 1980s throughout Latin America, 'disappearances' and torture were carried out systematically against all sectors. It was against this that the 'motherist' groups organized and feminist groups were able to provide a series of arguments highlighting the links between violence against 'the people' and violence against women in the domestic sphere. In Argentina, the presence of women in the streets and in the Plaza de Mayo signalled a different account of 'woman' in which political identity became tied to familial relations and in opposition to the overextension of the state into the 'private' sphere. The violence perpetrated by state agencies in the name of the people or the nation continues, as Jennifer Schirmer's chapter on the courage of Salvadoran and Guatemalan women shows so painfully and so clearly. Equally, as we write, women who are members of the feminist Flora Tristan organization in Peru have 'disappeared'. This organization takes its name from a Peruvian woman who, at the turn of the century, requested a divorce from her husband. Her attempts to change the legal framework within which women lived their lives exacted a high price; and so it is today that women activists throughout the continent engage in a politics with very high stakes and in which physical safety is not guaranteed.

The discourses surrounding the body are highly gendered, a point so often omitted from Foucault's own accounts of his 'bio-politics' (de Lauretis 1987; Diamond and Quinby 1988). In relation to sexuality and the technologies of power in play it is clear that Latin American states organize to promote heterosexuality and to police the sexuality of women through the generation of discourses on marriage, the family and fertility. In many states these discourses are allied with those of the Catholic Church in which birth control and the understanding of an active sexuality for women are not officially accepted. Nevertheless, the ways in which Church-based groups opened up spaces for women meant that issues of sexuality, the dynamics of male/female relations, birth control and reproductive rights were all brought into the debate if they were not yet able to be on the 'official' political agenda.

For example, the Christian feminist group Talitha Cumi (taking its name from an indigenous woman in Peruvian history) brings together concerns for women as valued, dignified and free persons:

> as feminists we understand the oppression suffered by women the world over. . . . From our perspective as Christian feminists, we believe that the theology of liberation can deepen its under-standing of the need women have to overcome their oppression and fully participate in every aspect of life.
>
> (Talitha Cumi 1984)

Similarly, Sister Zeca in Brazil has systematically questioned the role of women in the Church (Sarti 1989: 77). By this means politics is allied with the personal and becomes part of the technologies of self available to women throughout Latin America and a counter to the excessive 'policing' of state, Church and political parties determined on the production of docile female subjects. The Grandmothers of the Plaza de Mayo in Argentina express this practice in a distinct situation in their efforts to trace the children born in prison to incarcerated women, or the children detained along with relatives and subsequently placed under adoption (LADOC 1984). In these and similar organizations, women of varied ethnicities, classes, ages and histories have entered a terrain in which citizenship, nationhood and politics are negotiated and defined.

WOMEN AGAINST THE STATE: CLAIMING NATIONHOOD

Women in Latin America have confronted repressive state machineries as mothers seeking missing children and as wives claiming husbands who have 'disappeared'. In doing so they lay claim to their part in the nation and to their rights as citizens using the language of the state by reclaiming and thereby transforming it into a series of demands. It is generally recognized that the new social movements of Latin America represent an innovative form of politics which has recast political agendas and action (see Slater 1985; Ballón 1986; Jelin 1990; Redclift 1988). Women have been crucial to the human rights protests through-out Latin and Central America – the Plaza de Mayo mothers in Argentina, the Mutual Support Group in Guatemala and the Group of Relatives of the Detained-Disappeared in Chile, among others, are the best known outside their own countries, but there are also less well-known groups like the Mothers and Relatives of Those Tried by Military Justice in Uruguay (Rodriguez Villamil 1984). By organizing against state-initiated violence against families these women generated a process which had major implications for the ways in which feminist discourses have been articulated with civil liberties in recent decades, thereby generating specific indigenous feminisms. Jennifer Schirmer (1989) calls these groups 'motherist' groups in which women act as 'disobediant female subjects of the state' turning their powerlessness, as protected females within the family, on the state.

The most famous of these groups has been the Madres of Argentina; initially organized by the wife of a factory worker and the wife of a diplomat, they emphasized a cross-class alliance between women (Navarro 1989). They organized protests and petitions and found that they, too, joined 'the disappeared'. The state response was to recast the definition of the disappeared as the dead. Demands were made throughout the early 1980s and the period of the Malvinas (Falklands) War in Argentina which heralded a new nationalism in response to British aggression and new levels of repression in relation to the Madres. Since this time the group has split between those who were willing to negotiate with Alfonsín and those who were not. In 1986 an amnesty law was passed which effectively absolved the military in relation to 'the disappeared'; requiring that prosecutions for acts of repression should take place within sixty days. The issues remain unresolved.

In Guatemala the organization GAM has fuelled protests against the highest recorded numbers of disappeared, totalling some 38,000, perpetrated by the US-backed García death squads to suppress rural organizing and trade union developments. Women and men have attempted to reclaim a place in the nation by organizing resistance and denying the legitimacy of state action. In 1982 Ríos Mont massacred thousands of peasants, and finally in 1983 there was a *coup*. But, although Cerezo was democratically elected and made promises to stop the killings, it has not happened. GAM was started in 1984 by three women and, since its inception, it has involved many impoverished and indigenous women who have protested, organized petitions, taken on the judiciary and the political parties. They, in turn, have been terrorized, murdered and tortured alongside a government campaign to discredit the organization. But the protest has not stopped as Jennifer Schirmer's chapter in this volume attests, providing a reminder of the courage and the level of strategic acumen required for protest in Guatemala.

The Chilean organization the Group of Relatives of the Detained-Disappeared was started ten years before GAM, in 1974, by women who were ideologically leftists but who also sought to capitalize on the power

17

of 'mother symbols'. A hunger strike in 1978 brought an investigation which instigated the 'law of the presumed dead' and reparations for relatives. As in Argentina, the women refused this offer and increased their activities. Mass graves became a new site for protest and women chained themselves to the National Congress, while in 1984 women invaded the supreme court and unfurled banners demanding justice. An attempt to pursue indictments of the military and the police was overturned by the appeals court which voided all the indictments. The continuing process of negotiation over the role and voices of women in the transition to democracy is examined further in Catherine Boyle's chapter.

Generally, the above-mentioned groups present themselves as 'apolitical', emphasizing their familial roles as mothers and wives, and presenting their lives as totally disrupted by their losses which they are trying to recover. By so doing the women reassert the importance of family life and their roles within this in a putatively apolitical manner: the Madres explained that 'we don't defend ideologies; we defend life' (quoted in Jelin 1990: 204). By highlighting a 'natural' domestic role for themselves, the contradictions between their own agenda and that of feminists could be rejected. Nevertheless, they asked the men to leave their groups, using the image of mothers in need of protection against the state. This, in effect, politicized the family in ways which shifted it from the realm of the 'private' to that of the 'public'. Women involved have often been dismissed and discredited as 'crazy women'. Such groups are not always gender specific, however, and GAM is, in fact, the least gender-segregated of the organizations and the one in which ethnicity and 'otherization' is foregrounded. The exiled leader of the organization explains this in terms of the terror against which they struggle: 'the government doesn't distinguish between men, women, children, catechists, priests, professors, workers; it kills them all' (quoted in Schirmer 1989).

Action by motherist groups is predicated upon overcoming the private/public divide as it impresses upon women's lives. Thus, they have brought mothers in their domestic clothes to the centre of the public stage, symbolically protesting in the square outside the parliament in Buenos Aires. They also transformed the notion of 'disappearance' by refusing to accept that disappearance is death and turning it instead into a series of protests which insist upon remembering those who have 'disappeared', claiming by this a place in the nation's history for those the state wished to annihilate. The power of such motherist groups also arises from their ability to draw upon the feminine imagery of Catholicism against the state by evoking the image of the suffering mother and her sacrifice. To invoke this is one way in which to ensure political survival and personal safety for the members, although these

18

are not guaranteed. Equally, women sought in their protests to c
life in the face of death through the use of photos, candles and
which insist that the collective memory cannot be censored and that t.
like the disappeared for whom they grieve, are part of the nation.

'ENGENDERING DEMOCRACY'

The reference in our title is to Sonia Alvarez's pathbreaking study of
the developments of democracy in Brazil. From the Southern Cone to
the Andean Highlands, the moves to democratize have been as 'revolutionary' as the momentous changes in Europe and the Soviet Union. At
the beginning of the 1990s we see a Latin America in which in all the
Latin states there is a democratically elected president, except in the
increasingly isolated and weakened Cuba. Changes in political structures
have been allied with transformations in economic policy with shifts
towards more open economies under the auspices of international aid
agencies and banking institutions. These first tentative steps towards
democracy in the region have been noteworthy, not least for the role
that women have played in rekindling civil society and political identities
wiped out or gone underground during the repressive regimes that
marked the 1970s and persist still in Central America. As Jane
Jaquette comments: 'Ironically, military authoritarian rule, which intentionally depoliticized men and restricted the rights of "citizens" had the
unintended consequence of mobilizing marginal and normally apolitical
women' (Jaquette 1989: 5). This has, itself, led to debates concerning
the political role of women in the region. Commentators suggest that
women have often been at the forefront of struggles in the most difficult
periods of the repression of civil society, exemplified by the Madres
and civil rights groups discussed above. In many analyses, it is suggested
that once women open up the political stage, they retreat and do not
sustain the same level of political involvement. The complexities raised
by this are elaborated in Nikki Craske's and Yvonne Corcoran-Nantes'
discussions. It is an issue raised in relation to women's organizing
more generally and most recently framed within the context of
Maxine Molyneux's distinction between 'practical' and 'strategic' gender
interests.

The former term is used to define the ways in which women,
from the Andean Highlands to the Southern Cone states, organize
and have organized around their immediate, practical concerns of
collective consumption in the cities and the countryside throughout
Latin America. By contrast, strategic gender interests are, Molyneux
(1985) suggests, the shift away from the particular and the practical to
a more unifying ideology of women's interests on a general scale. While
these distinctions are useful in organizing what are commonsense

understandings of transformations in political strategies for women, and have been used critically by contributors to this volume and usefully by Sonia Alvarez (1990), the distinction is also problematic. It suggests that there is a simple dichotomy between 'practical' and 'strategic' gender interests which can be aligned with notions of the public and the private as spheres of interest for women; this, as we suggest, may be helpful for organizing commonsense but does not provide a theoretical base for understanding women as political subjects and actors. We would also want to suggest that it, too, has a universalizing quality which is located with a linear view of progress founded upon the post-Enlightenment account of movement towards a goal as part of the grand narrative of rational progress towards a better world (Boyne and Rattansi 1990). Such a meta-narrative suggests a hierarchical relationship between practical and strategic gender interests such that women, in order to progress, must move from one to the other. In addition, by reinforcing this sense of hierarchy it ignores the critique from feminism of the ideological basis of the distinction between public and private lives and it does not take into account the understanding from feminisms that the 'personal is political'. On the contrary, it tends to maintain the distinction between public and private, and between the personal and the political, the deconstruction of which has always been so central to feminist politics.

Instead, as the accounts in this book underline and the role of women in the civil rights protests in Guatemala, Argentina and Chile show, political identities are not fixed and the contexts within which they are mobilized are diverse. María-Pilar García Guadilla's chapter foregrounds this in relation to the ways in which environmental issues are constructed as 'political facts' through which women, and their concerns, become politicized. Teresa de Lauretis suggests an understanding of gender in terms of a 'subject engendered in the experiencing of race and class, as well as sexual relations; a subject, therefore, not unified but rather multiple, and not so much divided as contradicted' (de Lauretis 1987: 2). Our understanding, therefore, gives a central place to diversity and heterogeneity and to 'local narratives' through the conception of a multiplicity of sites in which women practise politics. There is no need for simple dichotomies between public and private, practical and strategic interests. Instead, these are recast in relation to the multiplicity of sites wherein women are engaged in power struggles, from the domestic sphere and the world of the household to the streets. Conceptualizing 'sites' invokes both the geography of space and the Foucauldian understanding of discursive space, or the social and power relations of specific contexts that have both formal recognition and informal negotiations as part of the ways in which they are constructed.

Sites may also serve to unify individuals who might otherwise be dispersed by the multiplicity of their interests. Thus, discussing the solidarity and democratization of everyday relations which arise in the Limeño shantytowns, Teresa Tovar suggests that 'the territorial identity is life and shared memory which arises from the barrio' (1986: 105). The barrio is a (physical) site in which people, particularly women, work on their common residence and by this means elaborate a tripartite political identity and agenda:

> constructing a collective identity that synthesizes and integrates the heterogeneous constituents of the barrios; legitimating the barrio and converting it into a collective protagonist in the popular movement . . .; and finally, turning the experience of barrio organization into a critique of centralization and exclusionary urban models.
>
> (ibid.)

Michel Foucault's writings, although often blind to the gendering of space, point to the ways in which particular 'technologies' of power and authority are grounded in and take their effects through the organization of space by social groups. As noted by one commentator, '[Foucault] sees spatial organization as an important part of social, economic and political strategies in particular contexts' (Driver 1985: 426). Teresa de Lauretis hints at the ways in which gendered identities, ideologies and practices are differentially constructed across space when she notes:

> the construction of gender goes on today through the various technologies of gender . . . and institutional discourses . . with power to control the field of social meaning and thus produce, promote and 'implant' representations of gender. But the terms of a different construction of gender also exist, in the margins of hegemonic discourses . . . these forms can also have a part in the construction of gender and their effects are rather at the 'local' level of resistances, in subjectivity and self-representation.
>
> (de Lauretis 1987: 18)

As de Lauretis is writing of Western societies her references to the 'margin' have a double resonance to our discussion of popular protest in Latin America where women are placed in the margins of the world economy, at the periphery of gendered power relations and in the underclass of formal power structures. In other words, it is in this particular 'place' in world politics in which Latin American low-income women, as seen in the chapters of this volume, organize, and through which they negotiate with other groups (whether these are dependent states, low-income men, and so on).

The physical sites in which women develop practices for protest are locations ascribed to women historically, the home/domestic sphere, or neighbourhood networks. These are geographical areas (or arenas) in which certain bodies are expected or even inscribed (see Ardener 1981), and in which other bodies are invisible or excluded. In other words, women are the 'expected' traditional inhabitors of the home in Latin America, and by contrast men are those found in parliament, army barracks and political party headquarters. Sites can thus become the centre for political action by the 'simple' presence of bodies in different sites: the example of the Madres in Plaza de Mayo springs immediately to mind here, and is undoubtedly a political act.

But sites also provide resources with which to act politically for men and women: public parks are crucial spaces for domestic servants allowed only a half day off a week; while isolated indigenous villages with associated minority languages in Guatemala permitted Rigoberta Menchú's political organization and the formation of political coalitions among peasant groups. The consolidation and elaboration of these sites and their associated resources is part of the process of politicization in Latin America. Resources and space for women's popular protest are provided by Church feminist groups, union organizations, and households, as shown in this volume.

A third element related to the physical-geographical sites in which gendered politics are initiated and elaborated is in highly spatially differentiated settings, where men and women have distinct lives both in relation to each other and in relation to men and women in other areas. Thus, gendered politics is different in Ecuador than in Peru, and within Ecuador gender patterns of voting differ between coastal Guayaquil and highland Quito (Prieto *et al.* 1987). However, state boundaries are often not the most significant factor, as the political, social and economic particularities of Latin America (the experience of conquest, the role of the Catholic Church, the Hispanic legal and political system) provide continuities across countries. Nevertheless, the chapters in this volume focus on the particularities of specific places and spaces, highlighting the diversity of women's participation in popular politics. The contrast between two areas of São Paulo, notable for its plethora of groups and agendas, described by Yvonne Corcoran-Nantes and Leda Machado in this volume, attest to this.

This, then, is a much more shifting account of politics which takes from Laclau and Mouffe (1985) the idea of frontiers and interfaces across and within specific sites. It is a way of understanding politics that is allied to an alternative, non-essentialist account of political identities. Equally, it is allied to the debates concerning what has been called postmodern politics in which writers like Boyne and Rattansi (1990) have discussed the problems of a Marxist inspired politics. Marxism privileged the grand narrative of class struggle, the defeat of exploitation and alienation through socialism and it is these themes that have fuelled political agendas throughout the twentieth century. But, the demise of organized class politics, or its appropriation by the state, has meant a move away from Leninist analysis towards and beyond a Gramscian account which foregrounds the politics of civil society and cultural leadership as explored by Bobbio (1989) and by Munck (1990). In analysing the Latin American situation, Munck reviews the role of Marxism in the continent's politics and arrives at the view that: 'Struggles are engaged in not to prove or disprove Marxist classics but because of a deep human will to surmount oppression' (Munck 1990: ix). Munck joins with Laclau and Mouffe in emphasizing the importance of plurality. He says, 'In accepting the plurality of political spaces we are implying the centrality of democratic struggles' and 'a dispersion of subject positions' (ibid.: 12). It is, therefore, no longer possible to engage in a politics of the singular, for example the class struggle or the women's movement, but we should understand instead the complexities of these struggles and the ways in which new terrain has been carved out for the political, expressed so often in the Latin American states through the impact of the new social movements (Vargas 1990). While these are indeed struggles around collective consumption they are also viewed as operating within discourses concerning the nation and the desire for the dispossessed population to have a place in the nation.

Equally, political ideologies/discourses have been organized around essentialist views of political subjects, the worker, 'women', but the ways in which the new politics fractures these imagined unities is itself profoundly emancipatory. Thus, a feminism organized around a unitary 'woman' in effect privileged certain voices and marginalized others, especially those of minority women, black women, indigenous women, women of the south – which comprise, of course, the majority of women. For example, women of African descent in Brazil have debated the issue of their alliance with feminists and have generated not one but a series of positions from distance and separation to issues-based alliances. The debates within feminism have fractured the unitary view of 'woman' and allowed for a conception of feminisms and of a politics built around both local and meta-narratives (Vargas 1990). Nancy Fraser and Linda J. Nicholson comment:

> whereas some women have common interests and face common enemies, such commonalities are by no means universal, rather they are interlaced with differences, even conflicts. This, then is a practice made up of overlapping alliances, no one circumscribable by an essential definition. One might best speak of it in the plural as the practice of 'feminisms'.
>
> (Fraser and Nicholson 1988: 102)

Thus, commonality and difference speak simultaneously and cannot be bound to a unitary view of woman or an overarching, trans-historical theory of patriarchy. Nowhere is this better understood than in the emerging democracies of South America where diverse women have organized through the base communities of the Roman Catholic Church as much as they have through feminisms.

POLITICAL IDENTITIES: GENDER, RACISM, CITIZENSHIP

The multiple sites of politics emphasized in our account is matched by the theorization of political identities as shifting. In the words of Ernesto Laclau and Chantal Mouffe (1985: 85) 'Unfixity has become the condition of every social identity'. Similarly, for Stuart Hall (1987), the coherent unified subject of earlier discourses is replaced by a shifting and contradictory collection of 'multiple selves' called forth by a multiplicity of discourses which has profound implications for political strategies and practice (Radcliffe 1990; Westwood 1991). The discursive and ideological formation of collectivities in Latin America has to be understood as part of the way in which discourses on liberalism and populism, for example, generate the collective subjects of politics. However, the political subjects currently being called forth as part of the new democracies are not without histories, ethnicities and cultures that place them in time and space in ways which the new politics seeks to acknowledge. Thus, the state and discourses on the nation seek to organize the category of 'citizen' which has been linked primarily to class in many of the Latin American states through party-based discourses around 'the workers', 'the peasants'. Women were usually marginalized, alongside specific ethnicities, because it was the commonsense of the region that women were a conservative force tied more to the Catholic Church and the home than the political sphere.

The histories of feminisms and women's political activism in Chile, Argentina, Brazil, Guatemala and across the continent tell a different story and one which the new 'democracies' can no longer ignore. A new account of citizens and democracy is one which acknowledges that 'citizens' are refracted through not only class identities but gender and ethnic identities as well. Thus, the ideological constructions of democracy require revisions in relation to discourses around racism and sexism if they are to be inclusive rather than exclusive, emancipatory rather than conservative.

Citizenship and democracy, therefore, are not reducible to one version of representative democracy located in ideologies of liberalism that generated a notion of citizenship tied historically to specific classes and masculinity. That democracy is contested terrain shows clearly in the cases of Peru and Colombia where the meanings of democracy are currently the source of violent clashes between competing ideologies (Bourque and Warren 1989). This foregrounds the understanding that there can be no taken-for-granted categories within politics and reiterates our view of a series of sites which extends the meanings of politics, democracy and citizenship. In fact, the politics of Latin America have shown this to be the case and it is evident in the chapters of the volume. In Brazil the women of São Paulo have used the Church as a means of generating political struggle. Similarly, in Nikki Craske's chapter the women of Guadalajara, famed for its religiosity, have reappropriated the Church as a space for political action outside the 'official' political party, thereby shifting the terrain of struggle. In both cases the struggles have in part related to issues around collective consumption but most especially to issues concerned with the body: bio-politics and health and welfare.

Sonia Alvarez's account of the opening up of political space in Brazil underlines the ways in which citizenship has been gendered by discourses around masculinities and the feminized world of home and motherhood. Clearly, the 'motherist' groups called this up in the encounters with the state machinery but, equally, the political parties offered women a specific place in relation to a familial ideology. Citizenship has been tied to labour-power and the male worker through discourses on class in the Latin states and women have had to fight for the basic right to vote which was not secured in Mexico, for example, until 1953. Equally, citizenship has been tied to specific ethnicities and exclusionary practices which have defined certain ethnic groups as 'aliens' as part of the process of 'Otherization'. Thus, even though the black populations of African descent in South America were imported to labour, labour was based not on the sale of labour-power but on enslavement which set them apart as workers and – because enslavement made them the property of another – did not make them 'freemen' who could be enfranchised in the earliest moves to citizenship. The fate of indigenous peoples tells a similar story.

State narratives generate representations (in relation to the ideological constructions) of masculinities and femininities as well as 'race', labour and communities. In this process there is an attempt to hegemonize political activism, as well as symbols and practices. However, popular social movements generate in turn their own counter-hegemonic accounts of the nation, gender and ethnicities which provide contestations to the state's projects and seek alternative forms of legitimation. This process is explored in Sarah Radcliffe's chapter in the specific context of Peru, by Yvonne Corcoran-Nantes in relation to Brazil and by Catherine Boyle in relation to representations of 'women', 'society' and gendered concerns produced by women in Chile in popular cultural forms. These issues point to the ways in which political identities are

25

formed through the cross-cutting discourses of popular culture very often used in turn by political parties as a means of mobilizing subordinate classes, as well as a means of resisting and subverting hegemonic narratives (Rowe and Schelling 1991). One, now famous, example of this is the figure of 'Superbarrio' who, dressed in a Superman suit, goes to the aid of the poor urban dwellers of Mexico City especially when evictions are in train. Superbarrio has shifted a male identity tied to the WASP-ish imperialism of US cultural life and reappropriated the power of Superman for the barrios. The figure of Superbarrio is a powerful symbol, NOT of Superman but of the barrio, the urban poor and their ability to organize themselves into the *Asamblea del Barrio*.

The contradictions explored in the popular cultures of the region can clearly be seen in relation to categories of 'motherhood', *machismo* and *marianismo*, which in commonsense understandings of Latin America are often foregrounded and reified as the archetypal characteristics of men and women (see Stevens 1973). Such views are exploited by political parties but they work on ideological constructions that are not simple unities but deeply contradictory phenomena, in which the contradictions can be exploited for counter-hegemonic gains, or which can allow alternative stories to be generated.

POWER BLOCS, POPULAR CLASSES AND POPULAR CULTURE

These relationships do not exist independently of the foregrounded issues of nation-state relations, racisms and economic development. The moves to democratize in many parts of the region are also related to the shift away from dependent capitalist economies with protected home markets and major debt problems towards more open economies undergoing fiscal and market-based reforms. Economics has often been perceived as the motor force behind political change in Latin America; for example, Guillermo O'Donnell's 1973 thesis on bureaucratic authoritarian regimes suggests that there were underlying economic reasons for the authoritarian regimes of the past. However, there is growing awareness that political development has its own dynamics and is not simply tied to economic change. In the early 1990s economic restructuring again looms large in the new scenario, with structural adjustment programmes encouraging privatization and reductions in tariff barriers; but it is not the whole story. There have been major political pressures, both from within and from without, to liberalize certain regimes (like Cuba and Nicaragua) while the USA continues to support repressive states in Central America. As always, it is never a single story. The recent changes, however, perhaps herald a move away from peripheral status in relation to the metropolitan core and capitalism as a global system.

Government attempts to liberalize economies by privatizing and integrating economic sectors more fully into world markets have major implications for the class politics and alliances within the region. The dominant class is fractionalized with regional power blocs versus capital cities in some countries. In others, as in Brazil, landowning families turned themselves into industrialists. These latter circumstances meant that economically the new power blocs had the power to define the national interest and their personal interests as one. Currently, the middle classes are placed uneasily within the state, with hyperinflation and debt crises reinforcing a sense of powerlessness. However, it is the marginalized populations which have suffered most in the recent dramatic political and economic transformations. For the urban and rural poor inflation is disastrous, affecting nutrition, health, workloads, intrafamilial relations and community organization. Caroline Moser's chapter shows how structural adjustment programmes have had a major impact at the household level, increasing the demands on women's time and thereby reducing their strength and time to dedicate to political movements.

Under these conditions the labour movement is in turmoil and has in some states been co-opted into a corporatist social formation, like Mexico, where the major political party and the unions seek to hegemonize political culture; other labour movements are devastated by economic crisis and shifting political priorities. In Argentina the trade union movement has a long history of militancy and it was one of the successes of Peronism that it managed to incorporate the trade union movement. The complexities of class formations and their political consequences relate to the fractionalization of classes via the formal/informal economy, the complexities of ethnicities, the urban and rural divide and gender divisions (see Alvarez 1990; Jaquette 1989). These complexities are rendered invisible via nationalisms which have developed narratives around the nation which have sought to integrate and exclude this diversity (see Collier 1979; O'Brien and Cammack 1985; O'Donnell et al. 1986; Malloy and Seligson 1987).

The current politics of Amazonia, for example, has ruptured the national story, forcefully bringing together the power of the multinationals against the nation and those who have consistently been marginalized in relation to the nation, state and economic powers. Equally, the indigenous peoples of Ecuador are fighting a similar, but less well-known, battle in relation to the multinational oil companies in order to secure their lands. Thus, the politics of the Latin American states involves not only national bourgeoisies, but international interests to which specific fractions of the bourgeoisie are wedded. This does not mean, however, that we need to revive a simple notion of the 'comprador' bourgeoisie. It is clear that there are contestations and accommodations between national and international interests and that these are currently being realigned in relation to the new economic policies and the moves towards democratization.

In bleaker times, when civil society was rendered invisible and driven underground, then the power of popular cultures through Central and Latin America to keep alive traditions, histories, and identities was crucial. The 1991 book by William Rowe and Vivien Schelling, while acknowledging the difficulty of writing a book on both the histories and contemporary manifestations of popular cultures in the region, is a vital resource and addition to our understanding of the ways in which when Victor Jarre sang in the stadium after the fall of Allende in Chile, he carried with him a long tradition of resistant cultural forms in song, music, poetry, dance and drama. But there has also been the attempt to give popular culture a hegemonic role in nation-building in which nationalisms have privileged specific ethnicities and accounts of racial democracy, a point with which we began this introduction.

The diversity of popular cultures is enormous, from Andean pipes and the Quechua language, dance and folkloric traditions to the urban samba and the cultures of African peoples in Brazil. Popular cultures as part of the contestations of politics have offered peoples from the Andean Highlands to the Southern Cone states an alternative account from the official orthodoxy, thereby offering them different subject positions – a factor which has contributed to the politics of identities. As we have sought to understand them within this volume, popular cultures offer contradictory accounts of the self and social positions as expressed for women. For example, the famous Latin American soap operas, the *telenovelas*, offer representations that exploit the contradictions of women's lives in the home, at work, in their relationships while presenting a politics of the personal.

This is also related to the idea that popular groups have historically- and place-specific 'repertoires' of actions with which to express grievances and articulate demands. As Susan Eckstein explains 'the Latin American repertoire has been shaped ... by dependence on foreign trade, technology, and capital, a bureaucratic centralist tradition, and a distinctive, Catholic-inspired worldview. ... The different modes of "popular" defiance ... are ... rooted in different traditions of protest' (Eckstein 1986: 10–11). As we see in the chapters of this volume, these repertoires are also differentiated by gender, as women develop different 'traditions of protest' to men, and as their experience of the features shaping Latin American political culture is in part (and to varying degrees) gendered.

The chapters in this volume speak to both diversity and similarity and to the contestations generated by 'disobediant' subjects throughout Latin America. They cannot be understood simply as representative, or as single country case studies, because all the contributions are given context by specific national boundaries within which there are very high levels of complexity and differentiation. What they offer us is a compelling vision of politics in relation to the issues raised in this first chapter and to which we return in the final chapter by Sarah Radcliffe.

2

THE SEEKING OF TRUTH AND THE GENDERING OF CONSCIOUSNESS

The CoMadres of El Salvador and the CONAVIGUA widows of Guatemala

Jennifer Schirmer

'One became strengthened with the discovery that one could do what one was doing – and many more things besides! Because if I think about who I was when I began, and who I am now, there is such a difference! For example, despite the fact that we aren't people with a lot of education, the situation obligated us to learn about national and international laws. Now there are CoMadres who cannot read or write but who can debate the socks off a lawyer about international law! And the CoMadre wins!'

(Alicia, a CoMadre directorate member)

'We didn't know what to do [after the massacres in 1982], we didn't cook, we didn't eat, our children just cried from hunger and pain. [But] we came to realize that we were left to carry the total responsibility for our family, to feed our children, apart from the great burden of suffering which we carry around in our hearts. . . . The first thing we had to conquer was our own fear.'

(Guadalupe, a CONAVIGUA directorate member)[1]

INTRODUCTION

How does a particular gendered consciousness and its discourse arise and become constituted? How does the experience of 'woman', 'mother', 'ladina', 'indigena', 'woman-Christian-activist' come to gain and shift in political meaning for certain women, and predominate in particular political and social settings? How is it that 'mothering' becomes so politicized that it goes far beyond its reproductive meanings, forcing women to see themselves as indigenous women and as peasant women?

30

Does a particular kind of state repression, and the resistance against it, collapse class and ethnic categories and create a gendered consciousness among women? Does the withholding of information and a censorship of memory by the state that asserts that mothers never had children create a movement by women which assumes a particularly stubborn seeking of the 'Truth'?

In short, do the motherist/widowist groups in Latin America represent a porous identity of womanist/feminist consciousness that arises from a particularly repressive, sexualized class and ethnic experience? And, finally, are feminist theoretical categories sufficient in helping us understand such transformations of and by women?

These are questions I came to ask myself while interviewing mothers and widows of the detained-disappeared in Latin America. This chapter begins with the CoMadres in El Salvador and the CONAVIGUA widows in Guatemala speaking of their transformative process of a constant re-evaluation and self-reflection involved in their gaining a gender consciousness under dramatically repressive conditions. In seeking to know who claims the 'Truth' about the disappearance of their relatives and in countering the repressive state's censorship of memory through disappearance, the CoMadres and CONAVIGUA widows have come to question the acceptance of violence against women, and have made a theoretical leap: they have connected their experience and analysis of political violence (disappearance and torture) to personal violence against women (rape and battering).

The chapter will end with an evaluation of feminist binary categories of women's actions, and ask how feminist theory might reformulate itself in light of such women's actions. By focusing on the political histories and demands of these groups, this chapter will look at several issues:

1 How for the most part *campesina* women gain a gendered consciousness and how this consciousness transforms their lives, with women, in some cases, moving beyond the 'motherist' paradigm to a larger objective of challenging traditional constructions of 'femininity', 'indigenism' and 'mothering' itself.

2 How this legacy of resistance is carried forward, not through formal organization, but through a readiness and willingness of daughters to continue to respond to threatening situations; and what the long-term consequences of these movements are for the next generation of women.

3 How alliances based on heretofore unknown cross-class and cross-ethnic relations arise and articulate themselves among these women.

31

COMMITTEE OF MOTHERS AND RELATIVES OF PRISONERS, THE DISAPPEARED AND THE POLITICALLY ASSASSINATED OF EL SALVADOR MONSIGNOR OSCAR ARNULFO ROMERO (COMADRES)

Political history

One of the founders of the CoMadres, Alicia, states:

'In the beginning of our struggle, it was an individual problem. But one began to discover that there were others in the same situation, and we realized we couldn't be isolated, that our struggle had to be collective. It had to be a group of people who participated in the same fight, in the same search, trying to discover the truth. . . . All of our committees [in Latin America] have arisen from this necessity: to seek out our relatives from the repression and to demand peace with justice!'

Every woman in the CoMadres has a disappeared, assassinated or jailed relative. The search by these mothers for their relatives in the morgues, the military barracks, the jails, the body dumps, began in April 1975, and by December 1977 a committee was formed under the direct auspices of the Archdiocese. Thus, the influence of Catholic liberal humanism upon the CoMadres cannot be overstated. Moreover, the peasant background of most of the women, and the kind of connections these women make with male and female workers at military and police institutions, reflects at once their social networking as well as their resourcefulness. Alicia again:

'There were only nine of us women then on Christmas Eve, 1977, when Mons. Romero and other priests met with us over dinner to help us officially form a committee. . . . He helped us transform ourselves from an individual to a collective search, to a collective struggle. Four days later, we presented to the President a letter which listed all the twenty-one names of the detained-disappeared who had been captured by the army. We demanded that he return to us all our relatives who were in the hands of the army. The President never replied, and by January 1978, the kidnapped, assassinated, disappeared increased – and our numbers [as CoMadres] were augmented to twenty-six mothers.'

The women clandestinely distributed flyers at the markets and during Sunday mass at the National Cathedral, and although the television and radio stations and newspaper offices were at the time under tight surveillance by the police and heavily-armed civilians, the CoMadres were able to evade them to deliver *comunicados*, *denuncias* and *campos*

Plate 2.1 June 1991 CoMadre rally in San Salvador on behalf of
campesinas who had been arrested and accused of 'subversion'

Photo: Mike Hutchison

pagados. They have occupied government offices, including those of the
OAS and the United Nations in San Salvador to telex their demands to
Washington and Geneva to put pressure on the Salvadoran government.
Today there is an estimated membership of 550 CoMadres (including
mothers of soldiers who have lost their children during the war and an
incalculable number who lend indirect support), with fifty full-time
CoMadres working in the office. They have staged sit-ins at the Ministry
of Justice, as well as in churches in different neighbourhoods in San
Salvador to educate people about the tactic of disappearance, and have
their own radio programme. Alicia:

> 'The radio programme covers everything: the negotiations, and
> political things, such as if there is a massacre, the *denuncia* is
> broadcast, if there is a capture, that, too, is broadcast. Everything
> that is news is broadcast.'

The mothers dress in black with white headscarves, Alicia explains:

> 'We borrowed from a march of mothers dressed in black during a
> march in 1922 to protest a massacre of male and female teachers.
> They went out into the street to protest the killing of their children.
> We borrowed our black dress from them. . . . Black signifies the
> condolences and affliction we carry for each person killed. And the

33

white headscarf represents the peace we are seeking – but it must be a peace with justice, not a peace with impunity! We also carry a red and white carnation: the red for the spilled blood, the white for the detained-disappeared and the green leaves, the hope for life. That is our complete dress.

Since we first went out into the street dressed like this in 1983, the army has never tried to disrupt our march, because it is not the same to capture someone who is dressed in civilian clothes to whom one can do anything, just anything, and one who is dressed in black, because everyone knows that that dress is of the Committee of Women [sic].'

Author: 'So, it is a form of protection?'

Alicia: 'Yes, the very military refer to us this way. The other day, we marched to the Treasury Police and stood outside shouting, "The National Police captured my son!" A policeman said to another, "Here come the women in black." And the other responded, "Ah, the CoMadres." So, they know, they know. The population also knows our work because they listen to the radio programmes. So, we are gaining a presence on many levels.'

Class differences

Of the nine original Co-Madres between 1975–7, most were *campesinas*, with 'background' support from housewives, teachers, a medical doctor, storekeepers, and a bookkeeper 'who provided us with some ideas and did some of the work, but weren't in the forefront. . . . We were also supported by a lawyer and a medical doctor with regard to political prisoners.'

To support themselves and their families, many of these women must sell fruit and vegetables in the markets, wash dishes in private homes and wash and iron clothes for the wealthy. At demonstrations, the women pass a bucket around to help mothers from the countryside pay for their bus transportation. As María Teresa relates:

'It was really difficult for many of us to travel to the city because we lived in distant provinces. Many of us from the countryside would get lost in San Salvador looking for different meeting places [before the CoMadres had an office] but little by little we became familiar with our surroundings, and the mothers in San Salvador would come to meet us at the bus stops. We were very impressed with the consciousness of the mothers who knew what to do and also how to help others in the same struggle.'[2]

I ask Alicia how this class mixture got on.

'One thing which motivated us a lot was that they [from the city] respected very much the opinion of the *madres campesinas*. They took what we said into account, and even though they were people who had studied, they didn't look down on our opinions. They would say, "Look, one must gather all the good ideas that the mothers have." And they respected us: there has always been a mutual respect among us, and I believe that that mutual respect has allowed us to exist until today. There is no devaluing our opinions because one or the other of us hasn't studied. The one thing that unites us is the problem of [finding] our relatives.'

While these peasant women publicly recognize that 'we are not intellectuals, and sometimes we have to do things four or five times before we get it right', there is at the same time a strong sense of ironic resentment as to how the government sees them. One CoMadre, América, says:

'To the government, we as women are incapable of understanding the political, social or economic situation. To them, we are only sexual objects for having children. We somehow cannot understand that a low salary for a factory worker means there is little hope of taking care of a family, we somehow are not aware of how *el patrón* doubly exploits us as women workers, we somehow are not living and seeing the effects of this war upon ourselves as mothers and women, and upon our children. What we are capable of understanding is that El Salvador is not a democracy, and that we are not afraid of speaking out! *Basta ya* to all the killing!'

Moreover, class resentment can act as a bond. I asked Alicia about the inclusion of mothers of soldiers among its members since 1984.

'The first thing that unites us, the mothers of soldiers and non-soldiers, is that we are all opposed to the army's forced recruitment. Most of the soldiers are poor sons of working people, sons of peasants. To work against forced recruitment is our common bond. Many of the soldiers are very young, and when they fight against the FMLN, they give up and become prisoners. The mothers then ask the FMLN to release their sons because it is not their fault that they must wear a uniform. In many instances, the FMLN has turned them over to the International Red Cross, which then turns them back to the army, where they are assassinated for being "guerrillas". For this reason, many mothers of soldiers have joined the CoMadres because of the unfair treatment their sons have received in the army. They were merely instruments of war, and when it is useful, they are assassinated.'[3]

Indeed, the CoMadres have worked with the mothers who have lost their sons in the army to gain the right to their sons' pensions. Many mothers have observed that when high-ranking officers are killed, the press covers it; but when a soldier dies, he isn't considered to be worth a line. Alicia relates:

'They are buried in a common grave and the family never receives any information about the cause of death. So the mothers see this inequality, and it has stirred them enough to join us. We treat the mother of a soldier in the same way that we treat the mother of a university student or a *campesina* – we give them the same kind of support.'

When asked if there have been any differences between indigenous women from the Maya-speaking areas, she responded that 'the communication among us is equal: we respect one another, and besides, many of these women also speak Spanish.'

Religious activism

Many of the CoMadres were involved in Christian-based community organizing in the 1970s, serving on Comunidades Eclesiales de Base (Cebs; Base Christian Community), leadership teams, executive boards of savings and loan cooperatives and community clinics. When these networks were targeted by the security forces, many of these women's relatives were killed or disappeared, even if their relatives had not been involved in Cebs. Alicia:

'Most of us at this time were from Christian-based communities . . . Mons. Romero would read our public letters out loud in his Sunday homily in the Cathedral so that the CoMadres became known nationally. One month before he was assassinated, I remember that Mons. Romero gave us his blessing by telling us, "Ah, women, you are the Marys of today. Mary spent a long time searching for her son, and you mothers are also walking along the same path that Mary walked. All of you are suffering the same loss, the same pain."'

For example, América Sosa is a CoMadre who worked in the Cebs. Her 14-year-old son, Juan, volunteered for the Green Cross. He was captured, tortured and beaten for ten months, and later gained political asylum in Mexico. América's husband, Joaquin, who did no political work, was abducted in 1981, and died fifteen days later due to his physical and psychological torture. Her son Joaquin, who is politically active in the Nongovernmental Human Rights Commission, was captured, accused of being an FMLN guerrilla, and imprisoned for one year – nearly dying from the torture. I ask América whether she has had previous political experience.

36

'I was working in Christian-based communities in the 1970s, which believed that people had to fight for their needs and organize to improve their standard of living. I was working for gaining a school in one community, and after five years of lotteries and dances to raise the funds, we were able to build one. In another colonia we worked on Sundays to construct huts, and we were able to build 240 houses in all. I remember this time fondly, talking with people, helping people. . . . The CoMadres then is a humanitarian organization infused with Christianity. Over time, we have fomented relations with both the Catholic Church as well as the Protestant one.'[4]

Having little previous political experience, the women initially depended upon one or two mothers to make decisions, gradually creating a network of *jefas* for each area of work. I ask Alicia how the CoMadres are organized.

'Until 1983, there was one person who, with all the ideas from everybody, was responsible for directing the work. She was Alicia Sayalendia who owned a small restaurant [*comedor*] and could talk easily with people, with the journalists, the delegations who arrived and thus, she was the most responsible for directing the work. . . . But in 1983, the directorate was voted on, and a responsible mother was named for each area of work. For example, there is the *jefa* in charge of investigations, another *jefa* in charge of the press and propaganda and they both travel to all parts of the country, taking photos [of corpses], taking testimonies [of witnesses] and all of that.'

The CoMadres's actions

Accompanied by a lawyer and a medical doctor, one of the CoMadres's first actions was to make an eight-day visit in a stationwagon as a small group to all the jails and all the military barracks around the country, asking after twenty-nine young men and women who had been captured and disappeared. In every place the soldiers would tell them, 'They haven't been captured, they aren't here. We don't know anything [about it].' When the group returned to San Salvador, they went directly to the National Police headquarters, to the First Infantry Brigade, and there they managed to find the son of one of the mothers:

Author: 'How did you find him?'
Alicia: 'We spoke with a woman who cooks for this Brigade, and she heard one of the soldiers mention his name, So, we went to [the chief officer] and said, "We know he is here." He denied it,

37

and we sent another letter to the President and published it in the newspapers. The boy was then transferred to another prison: he had been "disappeared" one year and three months, and we were finally allowed to visit him. He was very skinny, with a beard, long nails and in bad shape. We found a doctor for him through the International Red Cross, and he was released from prison in 1979. . . . And at the Treasury Police headquarters, the man who sweeps and cleans the garden recognized the photo we showed him, and we were able to locate another young man who had been captured. One month later, he was able to escape by slipping through the bars in his window, he had become so thin. And the Archbishop's residence gave him refuge and safe passage out of the country.'

In early 1978, the CoMadres occupied the Salvadoran Red Cross office and began a hunger strike, demanding the appearance of all those captured and disappeared before the press. By November 1978, they were marching

'together with almost all the priests of San Salvador. There were three blocks filled with priests, all dressed in white, who marched with us to the National Cathedral, completely silent, marching. A priest and a CoMadre were carrying a white banner that read "No More Kidnappings" [*Basta de capturas*].'

By January 1979, assassinations of priests was on the rise (in one community at least, five catechists were skinned alive). Based on these actions, the CoMadres began to take photos of the tortured and killed, and publish them in the newspapers with *denuncias*.

Clandestine cemeteries

When the Junta became government on 15 October, the CoMadres demanded that this government form an investigatory commission

'because we knew where secret mass graves existed. The Junta replied that no such secret mass graves existed, but we insisted and told them that we were so certain about this that we ourselves would form an investigative commission, and that we would select the members. And Marianella Villas [who was later assassinated],[5] a physician, a member of the Human Rights Commission, and five CoMadres all formed the Investigatory Commission, and we began our own search. We found forty-three places where bodies were buried, places where bodies had been buried for at least two years, other places where bodies had been buried only twenty days ago.'

It was during this search that a *compañera* who was representing the CoMadres in Mexico, Miriam Granados, found her husband. He had been left together with two other cadavers underneath the roots of several large trees. Alicia relates this incident:

'But one could see that his flesh had been destroyed with acid, because his bones were very, very white and the flesh was gone. But, in moving him, one could see that he still had live flesh on his face and that some of his bones still had blood. By this we knew that he had only been assassinated possibly no more than twenty-four hours before [we found him]. . . . Altogether three mothers found their children and it was very painful. One mother found both her nephew and son during this investigation in Cenate.

We found [in these mass graves] people without eyes, without tongues, without hands, people without heads. And in a place called La Montañita, in San Vicente, we found twenty craniums and skeletons. In the open were the animals – the vultures, the pigs, the dogs – who had torn apart the cadavers and were eating them. But from the craniums we knew that there were twenty cadavers. . . . But we didn't have the experience to know then to be careful with the skeletons, so we put the bones in aluminium boxes and brought them in to the Ministry of Justice and the Public Ministry so they could carry out a more thorough investigation. But they did nothing with all of this.'

Author: 'Did you continue taking photos?'

Alicia: 'Yes, these are now in Mexico, because all that we had [in the office] the National Police carried away.'

Threats: the state's attempt to censor memory

The costs of such defiance have been high for the CoMadres. 'It was in May 1978,' Alicia relates, 'that we received our first threat. [The letter] said "Give up [your work]. Stop going about asking questions, publishing things in newspapers if you don't want to be disappeared, one by one."' By August, a threat by Major Roberto D'Aubuisson, head of the death squads, was published in the paper which read 'Those phantom women [*fantasmas*], one by one they will be decapitated.' This was repeated by the President, Col. Carlos Romero.

This reference by the government to the women's phantomness effectively demonstrates the confused attitude of the government towards these women: on the one hand, they are a phantom organization, they don't exist, nor did the disappeared ever exist – i.e. it is an invention of these women; on the other hand, they are such a threat, they are publicly speaking the truth and challenging the government, that they must be killed. One might ask, how does one kill ghosts? Why is one so concerned with lies? Clearly, these repressive actions are an attempt to censor the memory of 'the public' of 'disappearance' and 'capture'.

But the CoMadres persevered, as Alicia says:

'With this first threat, we went to the Red Cross for protection, and we stayed there for three months and when we left, the newly formed Non-governmental Human Rights Commission provided us with space where we could meet. We were now a larger group, with women coming in from all over the country. One such group was from Metapan Lake, wives of fishermen who had been disappeared. They went with us to the ministries, asking them to investigate the whereabouts of these men, but it was useless.

We tried to legally petition the Minister of Justice and the President, but we were told 'Women, your relatives are in such-and-such a place [outside of the country].' Or they would tell us, 'Your son is in the mountains fighting as a guerrilla.' But these legal actions didn't evoke any response, so we thought about public marches in the streets two or three times a week and sitting on the steps of the Cathedral every Friday. We would announce these marches with flyers we would hand out clandestinely at Sunday mass at the Cathedral and in the marketplaces.'

Their first office, shared with the Non-governmental Human Rights Commission, was under surveillance by military forces 'circling the place but never entering'. It was bombed at 4am in 1980 and partially destroyed (there were no victims). There have since been bombings of the CoMadres offices in 1981, 1986, 1987 and 1989 which destroyed the infrastructure, and wounded *compañeras* both physically and psychologically. At this time, as Alicia recounts, a state of siege was in effect and

'nobody, nobody went out into the streets, there was this total wall of silence in which no one said anything, no one pronounced a word [about the situation], while the large military operations in the countryside were taking place, burning houses, killing people. It was very tense. And the number of mothers rose dramatically: in just one place, we gained 700 mothers. For this reason, the committee consists of mothers who don't know how to read or write because they are peasants from the countryside as well as peasants from the city.'

The most destructive bombing was on 31 October 1989, in which six CoMadres, a North American and a 4-year-old boy were seriously wounded. Two weeks later, on 15 November, the Treasury Police arrived in the middle of the night at a house of a CoMadre and captured four CoMadres, a relative and two North Americans. The following day, the National Police arrived at the office and carried off everything of value, including $500,000 in medicines that were to be distributed in the hospitals and public clinics after the November Offensive. CoMadre María Teresa says, 'After that, no one wanted to rent us a place, and we had no place to go until a group of feminists in the Norwegian Labour Party donated money to us to buy a house.'

After the November 1989 Offensive, the government stepped up the repression and, on 23 December, the CoMadres held a press conference upon opening their new office, breaking the fabric of terror established by the ARENA government.

Despite the shift by the security forces toward more 'selective repression' by the mid- to late 1980s, street demonstrations were still repressed. For example, in January 1989, a peaceful march to the San Carlos Garrison in protest at forced recruitment was harshly intervened upon by First Infantry Brigade soldiers who fired shots, threw teargas grenades into the crowd and abducted thirty people, including several of the CoMadres (CoMadres Human Rights Alert, Washington, 9 January 1989).

CoMadre Alicia recalls:

'We were being threatened by the Treasury Police. The death squads captured a *compañera* Maria Ophelia Lopez. She was detained, tortured and raped. She was tied down by her hands and feet and burned with cigarettes. Every time they showed her a photo of a different CoMadre and asked if she knew them, when she replied no, they would torture her. She was released eighteen days later on the highway to Santa Ana, outside of San Salvador. Some peasants found her and brought her back to the capital. Her face was badly bruised, her teeth were broken, her thumbs were tied behind her back.'

In 1982, three mothers, aged 46, 52 and 76, were captured by the death squads on their way to visit the men's prison. They were detained in the offices of the National Police and violently tortured. They were taken to the Devil's Gate (a body dump) and told that if they did not comply, or agree to their accusations or answer their questions in the way they wanted them to, they would be decapit..ted. The National Police released two of the women, and imprison⟨ ↳ the third (the 52-year-old) in the women's prison. (Her husband ai.d four sons had been disappeared; 'they eventually rounded up the two daughters'.) She was released in the 1983 amnesty, and she fled to Canada with three of her children. 'She has never known what happened to the rest of her family.'

In 1983, another CoMadre was assassinated with her husband in her house by death squads. She had been raising her daughter's two children after she had been disappeared together with her husband. Also on 20 August 1983, another *compañera*, María Elena de Recino, was captured together with her 13-year-old daughter, Ana Yanira Recinos, and a lawyer who often helped the CoMadres, América Fernanda Perton. They were captured by the National Police at María's home where they were planning actions to help free unionists who were being held prisoner (one of whom was María's husband), put into a large truck and taken to Treasury Police headquarters. 'We have made many requests for information and staged actions in the street, but they remain "disappeared" even today', Alicia stated.

In July 1989, three CoMadres were captured by the National Police while sitting in front of the National Cathedral protesting the repressive Anti-Terrorist Law of the ARENA party.[6] Once again, in 1989, the Treasury Police invaded the CoMadres's office and abducted several mothers, some of whom were released two days later. Another CoMadre was captured by security forces on 29 January 1991, when she accompanied a boy who had been wounded by the bombings to the hospital. She had accompanied a group of US congressmen who were in El Salvador a few days before. She was tortured by psychological pressure, beatings and repeated rape. They threw her down on the floor, threw water on her and gave her electric shocks. They kept her naked in a cold room, with her hands tied above her head, They also threw her into the trunk of a car and asked her where she would like to go: to *El Playón* or the Devil's Gate (*Puerta del Diablo*) (both are body dumps). With her hands tied and mouth gagged, they gave her electric shocks, and afterwards threw her into a dark room. She could only feel that the floor was damp, and touched a damp spot with her finger. When she raised her finger to her nose to smell, it smelled of blood. And feeling around, she could detect shattered bones, and picked up a human finger. She thought they were going to kill her. But the CoMadres waged a campaign on her behalf, contacting the US congressmen, who kept calling the Salvadoran Embassy in Washington. This helped gain her release on 5 February. 'But what she saw in that room and what they did to her is to witness what goes on in the military garrisons', Alicia said forcefully.

After each rape, she said she was forced to curl up and be washed with hoses so as not to leave any evidence. But when she was released, the CoMadres took her to a doctor who examined her and wrote in his report that she had multiple lacerations from being raped. Alicia, crying, says:

'But she remains in poor psychological condition. She is very emotionally debilitated from this experience, because they threatened to capture the other children in the family. And they threatened to capture me, because you see, she is my daughter. The experience has left its lesions, as it does for many of our *compañeras*.'

Author: 'Do you have anyone to provide psychological and medical attention to the *compañeras*?'

Alicia: 'No, not yet, but we have spoken with several institutions that can provide this for us. Because of the costs, we give priority to the more delicate problems, and leave the others. . . . But this means that we neglect ourselves a little because of lack of economic resources.'

Author: 'Do you think that the use of rape against yourselves is well planned by the security forces? That is, since they can't stop your activities, they plan to destroy you psychologically?'

Alicia: 'You mean destroy us using another method?'

Author: 'Yes, torture combined with rape.'

Alicia: 'Yes, that is planned. They know that one way you can emotionally injure a woman is by raping her. For us, it is something very delicate and, in this sense, they know this perfectly well, so they use this method against us. And since it is the worst thing that can happen to us, it could effectively destroy us. Here in El Salvador, there are many cases when a married woman is captured and is tortured and when her husband finds out about it, he leaves her. They [security forces] know this, and that's why they use it against us. There are also cases in which there is more understanding by the husband.'

The CoMadres are clear about the government's intentions – as well as its failures. Alicia continues:

'They use rape because the government and army have not been able to make our organization disappear. The same military types say, "To capture a CoMadre is to fall on top of an anthill." They themselves say that. Because we have international support, and when there is a capture, international pressure mounts. So, they capture people, but these captures initiate international campaigns. Yet, these captures are planned out. And since they can't make us disappear, they use other methods, such as internal beatings, and forcing [by rape] bacteria into our vaginas. Someone who is sick cannot work, right? It's all part of the repression against us. It's part of the politics of wanting us to disappear, of wanting to destabilize us. But my daughter wants to continue working, because what she was doing when she was captured – giving medical attention to someone wounded in the bombing – is something our government should be doing. But instead, the police see this kind of action [on our part] as dangerous and carry

43

out a capture. We had been under surveillance for several days, but we never thought it would be carried to such an extreme. But they are capable of anything; that has been proven. And so are we.'

In all, it is estimated that more than forty CoMadres have been captured, raped and tortured, two who remain disappeared and one who was assassinated. Alicia says:

'Apart from [the pain] of having lost a child, apart from this, there isn't one of us in the committee who hasn't been captured, raped and tortured. . . . With all this surveillance and persecution, to inculcate terror within us, perhaps with all these things they have done to us, it has made us lose our fear. Even though the military is present [on the street], we walk past them as though they were not there.'

It is clear that governmental security forces use such repressive tactics against the CoMadres, believing that if the formal structures of an oppositional organization – the office, and even the leadership – are destroyed, then that organization will fall apart. What the security forces apparently have not come to realize is that the CoMadres, as other popular organizations in El Salvador, arose from and are able to survive a lack of formal structure; that is their strength, and that is the strength of many women's popular organizations throughout the world. Because so many women are marginalized from the formal political settings of political party, trade union and Church hierarchy, they learn to create their own flexible forms of organization outside the rules of the game. So that when the government approaches such an organization to destroy it on the basis of its understanding of politics, it is clearly frustrated that it cannot destroy these women's will to continue to organize and mobilize, despite the repeated bombings, captures, tortures and rapes. In fact, rather than hide from fear, these women seek out the public spaces to coordinate their work, as María Teresa recounts, 'Although we were left without an office [after the bombings], this does not mean to say that the CoMadres fell apart. We would meet and do our coordinating work in the parks and streets.' Alicia also recounts:

'But it was in 1981 when a bomb was thrown into our office and left us with absolutely nothing. What we had repaired from the first bomb, they returned to carry away or destroy. We had to meet at a concrete table under a mango tree within the grounds of the archdiocese, where we would receive the many *denuncias* from the internal refugees fleeing the bombing of their villages and we would feed and clothe them.'

44

The CoMadres's projects

The CoMadres's projects have arisen out of a recognition of the need to provide food and clothing to refugees and prisoners. This work has encouraged women who were unaware of the CoMadres to join; it has also, in turn, served as a form of consciousness-raising among the CoMadres themselves, allowing them to see first hand the effects of the operations of security forces, as CoMadre María Teresa relates:

'My husband was a worker and was captured by the army. In my own five days of not knowing what had happened to my husband, I became aware of the suffering of the CoMadres. I later found out that he had been held *incommunicado* by the National Police. He was then transferred to a prison, where I met the CoMadres who were coming with food for the prisoners. That is how I became familiar with their work. And I would see on Sundays how many women would come to the prison with the hope of finding their husband or son there. I also became aware of the different kinds of tortures that they apply. I joined the CoMadres then in 1978; I had never been involved in any political work nor did I know if the CoMadres were a political or humanitarian group. All I knew was that they understood my suffering.'

In many of these Christian-based communities affected by violence that the CoMadres visited, they saw an increasing number of orphans who needed to be fed and cared for:

Alicia: 'We gathered together about a hundred children in one house we rented in Colonia Santa Lucia where they were fed and clothed and given all the necessary attention by five *madres* and support from the community. But we ran into some bad luck because the army – more specifically, the air force – found the group of children, discovered they were orphans of assassinated parents, and came into the house and took away eighty-four of the children – all of whom later appeared dead at El Playón [the body dump near San Salvador]. The youngest was 4 years old and the oldest was 9. It was such a painful loss for us. . . . They only left us sixteen children, and these we took to several private homes to protect them. And there were many more, but we had learned not to house them in one place.'

The CoMadres have established daycare centres throughout the city and countryside for working women:

'We believe in not just speaking but also in doing, and no one believed it could be done, because the daycare centres here have only been run by the government with uniformed children, hair

45

cut to the same length, standing in orderly lines. . . . And what is it we want to do with these daycare centres? Do we want to just continue with the same attitudes? No, we must change them: the fathers must begin to change their thinking about the rights of the child, and about the mother! So, besides the daycare centres, we need to have night classes for fathers so they can learn to share in the responsibility of childrearing – classes that could be called "The Education of the Child". Today we are planting the little seed for the future, for a better childhood for our children, for a better approach by the fathers. The most important work for us is to change attitudes. We work with this vision of the future; it gives us hope.'

The CoMadres have also established a health clinic which averages twenty-five to forty patients a day each Monday and Saturday, at which they provide free medicine. They have also organized sewing workshops for free, to provide training to families without jobs or skills.

Gaining new ways of knowing and speaking

'Because the problem is so pressing', the CoMadres began to acquire new ways of knowing. At times, though, according to Alicia:

'it seems like we're doing nothing at all. But one transcends this feeling, right? Because in our case, you can't tell us "Look, what you're saying is a lie." Nor can they silence us, for since the moment that there has been neither a response [to our demands] nor a clarification, our fight has been legitimate. It is legitimate, because we have the right as human beings to know, and as mothers, we also have the right to know the whereabouts of our child who was captured and disappeared.'

Another CoMadre, María Teresa, explained it this way:

'What I mean when I say that we are not a political organization is that we are not a political party. It has been by living through and enduring this war that we have developed a political activism. With much difficulty, we had to write things ten times before they turned out correctly, but this experience has taught us how to be political activists. We had to speak out, so it just happened. And you can see from our publications and *denuncias* in the papers that we are not lawyers or professionals; what comes out is pure mother. We make a lot of grammatical mistakes, but we feel what we do to be valid in a country without a god, as we say. Just listening to different things on the radio, we now pay more attention, we are able to comprehend what is going on, and to develop our own way of speaking about it.'

I asked CoMadre Cecilia what she thought of the feminist movement:

'I don't have any problems with the word "feminist"; I understand that a feminist movement fights to regain possession of rights of women as a gender, right? To have the right to the same rights as men: to work, to health, to study, to participate politically if she wants. Although, I will say that there are women who fight against the man, and feminism for us is not that. For us, it means an equality of rights, all of this work we do together to create an equal level [of conditions] for women and men. Not to separate men from women; that's not our vision. But to walk together on equal terms. That, I think, is where the difficulty lies [with some feminists].

We [CoMadres] conducted a study internal to our organization and we asked the *compañeras* about their relationships at home. We discovered that many women got married but they didn't know why. And we had meetings that lasted into the night, until dawn, discussing this. Often, the woman is abused by her lifelong companion, and she hasn't discovered that she doesn't have to accept this abuse. So, we have seen within our own organization, the significance of the fight for women's rights as we've discussed each problem with every *compañera*.

And we need to practise these rights for women in El Salvador because there is no education for women. Most women here believe that they were born to keep house, have children and wait on their husband. They don't know that women have much more capability. . . . But I also believe that the struggle by feminists, in order to be an effective struggle, must begin by educating the opposite sex — the man, because if he's not educated about the women's right to excel and the right to be well treated on equal terms, then you're always going to fail.'

Author: 'So, would you say that you as members of the CoMadres have learned a lot these past fourteen years?'

Cecilia: 'Yes, we've gained more consciousness and more under-standing. More understanding because only now do I understand what the fight for human rights is all about. It is the fight for the rights of the workers, for the rights of women — before, I didn't know this. And I have discovered this: human rights are not just about defending life. Defending life is a basic element, certainly, and the most important, but surrounding this right there are many more things: the right to study, the right to health, the right to shelter, the right to culture, the right to recreation. So, yes, I have developed a lot in my consciousness since I've been active in the CoMadres.'

The legacy of struggle: mothers and daughters

I ask Alicia whether the next generation will continue the struggle of the CoMadres, and she connects the issue to changing attitudes and the need to continue struggling within a democratic transition:

'Yes, because there have been many *compañeras* who, for reasons of sickness, have died and their daughters have taken over the work of their mothers. This is quite natural. . . . But we have a vision of what our people need. Why would we want to create more widows? We believe that the project to investigate the disappeared is going to remain viable in El Salvador. It is viable and it will continue being viable for many more years, because independent of all the governments we've had and are going to have, an investigation must be made about all the people who have disappeared. Without the humanitarian organisms, without trying to be hegemonists, but simply with the truth and with the reality which has become our own, we will be able to say, "*Señores*, provide us an accounting of these, these and these relatives we have lost and who are listed as disappeared, and that this guy and that guy were involved." We will also seek indictments and sentencings. But we will not ask that they be disappeared as well, because that is not our objective.

Clearly, one must prepare the people [for this change]: especially in terms of a change of attitude. They are now projecting a new politics of democracy, and these people must have another way of thinking. Can you imagine if we were to continue creating people who were going to torture us? Certainly not. This is a job of education and attitude that the CoMadres are planning for the future.

We aren't going to stop our work because a new [democratic] government comes in; we still have to fight for the disappeared, we still have to struggle against impunity. We still have to seek indictments and sentencings.'

Another CoMadre, América, puts it simply, 'Our committee will continue until we win peace with justice.'

Conclusion

It is clear that the CoMadres recognize a strong collective memory of social ideals in El Salvador, ideals to be taken up by the next generation. Their sense of having a historical responsibility to prevent the impunity with which security forces violate human rights transcends traditional individualism. Rather, it stands apart as a form of collective citizenry in which the present generation owes it to the next generation to prevent the horrific experience of the 1980s in El Salvador from reoccurring. Such a citizenry entails the right to live without the threat of such impunity; it also entails the obligation to pass on the collective memory of such impunity to one's daughters and sons so that the past will not be forgotten – or repeated. This added sense of responsibility for their future children and grandchildren, it could be said, reflects a sense not only of collective citizenry but also of political motherhood – and potentially fatherhood in their future vision – in which they believe that there can be no national security without social justice.

I ask Alicia, one of the original founders of the CoMadres,:

'Given all of the violence you have experienced, and the failure of the courts to find any of your relatives, how do you maintain your faith in the rule of law? How do you maintain your hope?'

Alicia: 'There are two elements which are very real to us: one, the need to know the truth; and two, that we seek ways to change the situation completely so that the next generations won't confront what we have had to confront. We cannot permit them to live through what we have had to live through. And in this sense, yes, we have much faith. Also, the spirit of Monsignor Romero is always with us; he gives us much strength, much willpower. It is a faith that we keep alive.

We know that what we have done is just, and we know we have contributed something, although it seems so small. . . . It is going to be a society which will live without impunity – that is what we are looking for. We ourselves aren't going to see this new society, but it will occur. As Mons. Romero said, "They can kill my body, but not my ideals." Ideals never die. We will keep them alive so they may be taken up by other generations, so they can put them into practice. We must succeed in preventing impunity [of the security forces for human rights violations], because what we have had to experience has been sufficiently difficult, and we cannot allow the next generations to experience the same thing.'

THE NATIONAL COORDINATOR OF WIDOWS OF GUATEMALA (CONAVIGUA)[7]

'[In the highlands after the massacres], we watched the widows: how many there were, how they obtained their food, who gave them food, and where the orphans were [located], and who took care of them [in order to determine who and where the subversives were].'

Guatemalan Defence Minister General Gramajo
(author interview, March 1991)[8]

Political history

CONAVIGUA, the National Coordinator of Widows of Guatemala, is a group comprised mostly of indigenous, rural women widowed by the political repression in the early 1980s, or by illness and the brutal conditions on the *fincas* (sugar, cotton and coffee plantations), as well as other women – indigenous or non-indigenous, urban or rural – who wish to join them in their struggle for women's rights. As one Council member, Guadalupe,[9] states:

> 'Our organization is open to all women because our situation as women is that our participation in society has yet to be recognized; we do not all start out equally. At the moment, most of us are indigenous women, most of us are widows. Our struggle then is to see that women are respected and given our rights, and to do this, we must unite ourselves against the repression that we all face.'[10]

CONAVIGUA was formed by widows from several villages after meeting through Church activities, while making *denuncias* to the military or while working on the *fincas*. Initially, CONAVIGUA was formed to address their most immediate needs for food, medicine, clothing and housing. On 14 May 1988, in the highland province of El Quiche (the site of ferocious counterinsurgency operations between 1981–4), the women organized their first public religious activity as a form of protest, as Maria relates:

> 'This was only four days after national Mother's Day. But for us, thousands of *mamás* don't know where their children are, and thousands of wives don't know where their husbands are. We refused to recognize Mothers' Day, so the priest there helped us create our own day for remembrance. . . . But what happened over the next couple of days after our protest? The military authorities called a counterdemonstration against the priest, and came looking for him, but he had already left the parish in fear. From that experience, we could see that [the military authorities] wouldn't let us organize! Wouldn't let us make public denunciations and speak out about our suffering! We came to realize that only by uniting could we achieve our goal. If we couldn't make denunciations, what were we going to do? We simply had to speak out to say that the repression left 60,000 of us widows in Guatemala.[11] So, we have been doubly affected: now we are widows with children trying to earn a living for our families.'

These indigenous peasant widows had to learn to till the soil, to gather firewood, and to leave their homes to sell their products in the marketplace 'without knowing how to do this'. Some of them have small plots of land to till, yet the prices of corn and fertilizer have risen beyond their meagre means, making self-sufficiency even more difficult to attain. Other women left with their children for the south coast to harvest cotton, sugar cane and coffee to earn enough to eat 'to find one more tortilla for our children'.

Often, the *finquero*, knowing that the indigenous worker cannot read, cheats with the scale, and thus pays the worker less than Q1 (50 cents) a day. 'We know how heavy each sack is: we're the ones who must carry it on our backs! They always pretend it weighs less and, instead of being given our dues, we're robbed' (Lucia quoted in Frankel 1990).

The widows are also aware of gender discrimination, as María recounts:

'In some cases, [the *finquero*] gave us the same or even more work to do than the men, but they paid us less. . . . On the fincas, if one brings along one's children, then [the *finquero*] says that women only work to be able to feed their children, and so pay us less. Women are not valued for ourselves, even though we do all the work. So, we must organize ourselves and do things for ourselves, for I can tell you, no one will do them for us!'

Two years into the increasingly disappointing civilian regime of President Vinicio Cerezo, on 10 to 12 September 1988, the First National Assembly of Widowed Women was held at which a National Council was elected by majority vote. As one member of the directorate, María, relates:

'We had already been forming committees by way of institutions, such as the Church, in communities for several years, and these groups were evolving over time. Suffering brought us all together; but in the midst of such suffering, there is also deception.'
Author: 'Why did you organize yourselves in 1988 then?'
Maria: 'Because the government of [President] Cerezo had come to our communities, promising seeds, tools, loans to pay debts – a whole bunch of things – and some of our communities formed committees to request support for their teachers and health workers [*promotores de salud*]. But once they arrived at the government offices, they realized it was all a trick. [The government and military people] threw our people's demands into the wastebasket, and insulted the women, calling us "bad women". And we couldn't respond in Spanish.'
Author: 'Why do you think they called you "bad women"?'
Maria: 'Because we had left our homes to demand help from the [government and military] authorities. We told them what we wanted, but, you know, there isn't any freedom of expression in Guatemala.'

Religious activism

Several of the nine directorate members have been previously active in base-community (Cebs) activities of their local Catholic church, called 'Acción Católica'. However, several of the women explained that although they were part of this religious organization, they really didn't know how to organize as a group, how to work together.

> Guadalupe: 'It has been through our work here at CONAVIGUA that we have learned so much. For example, I didn't know how to speak Spanish three years ago, and now listen to me! No one on the directorate has been to the university, or to any school. We have all come from the country and thus came out of the school of the kitchen and family to attend our meetings. Only one member has learned from textbooks, because she completed the first six years of schooling. . . . So, you see, if we [in CONAVIGUA] hadn't organized, we wouldn't have come to learn about all this.'
>
> Author: 'But because many of you speak distinct [Mayan] dialects, how do you communicate?'
>
> Guadalupe: 'No, we can more or less understand one another between dialects. There are many words that are similar.'

Of the approximately 9,000 members, there are only twenty-five or thirty who know how to read and write.

CONAVIGUA's goals

Their motto in 1988, 'For the dignity and unity of women, [we are] present in the struggle of the people', suggests their objectives:

> 'for women to participate, for our opinions to be taken seriously, for our dignity and rights as women to be respected. And how is that accomplished? By participating in our social, cultural and religious actions, and in larger marches and demonstrations called by UASP [Unidad de Acción Sindical y Popular], we show that women must also participate.'[12]

Moreover, they demand that their dignity be respected, as women and as widows who are struggling against the many abuses present in their lives, to fight against the abuse, rape and exploitation they suffer at the hands of the military.

Another goal is to stop the forced recruitment of these women's sons, as widow Juana relates (see Figure 2.1),

> 'we don't want them taken off to military bases for forced labour, for the civil patrols. They are taken off like a rock, like an animal without feelings. We receive nothing from the government or the

Figure 2.1 Flyer handed out by CONAVIGUA and CERJ demonstrators demanding an end to forced recruitment by the army, Santa Cruz del Quiché, June 1991

army to raise our sons – no food, no education – why should our sons, in turn, serve this government or army?'[13]

Another of CONAVIGUA's aims is to obtain assistance and some sort of compensation for women whose husbands were killed during the counterinsurgency campaigns of the 1980s. Their activities raise the consciousness about the wretchedly poor conditions of 87 per cent of the Guatemalan population, most of whom are indigenous, who can afford to eat only tortillas with salt, and who cannot meet basic nutritional needs.[14]

Furthermore, they demand respect for human rights, prosecution of those responsible for repression, and permission to retrieve the remains of relatives from clandestine cemeteries.

CONAVIGUA's activities

CONAVIGUA's activities seek better education and better healthcare for their children. They have established small weaving and agricultural cooperatives, such as a maize mill and rabbit farming.

One of the deepest concerns of CONAVIGUA is the physical and psychological health of indigenous children, 50 per cent of whom die before the age of five in Guatemala. The CONAVIGUA Statement of 18–19 July 1991 states:

> Many of our children die of hunger, malnutrition and overwork. Many have been jailed together with their mothers. But perhaps the most harmful is what our children have seen due to the repression: seeing their house destroyed, and suffering the great trauma of seeing with their own eyes the torture, assassination or kidnapping of their parents. Many maintain the uncertainty that if their parents disappeared, that they will return some day. The psychological damage which has been caused perhaps can never be repaired.
>
> (CONAVIGUA 1991b)

Literacy for women

With Spanish literacy classes, women learn the language of bureaucracy, law and power as a way both to make their demands known and to protect themselves. Learning to read and write can be a means to regain a sense of control over one's life and one's body integrity. María states:

> 'We've always lived in ignorance, not knowing if we can really expect respect or have the right to protest such a thing as rape.

54

Now [that] we have the experience of becoming organized, this becomes possible [to consider]. Men frequently misunderstand what we're doing, assuming we only get together to hunt for husbands. They hassle us and we show them [that] we have the right to complain – and they're even less . . . pleased! If an offence is committed against a woman, we now know how to bring the man before the authorities. That teaches the men to respect us as human beings.'

Moreover, members of CONAVIGUA attend meetings, talks and lectures covering such topics as the Constitution, Maria says, the right:

'to organize ourselves, women's rights, freedom of expression. These have helped us very much. I, for instance, didn't know that there were laws in our country, that there existed rights or a constitution. Because we have never seen these laws fulfilled or rights respected, all of this means nothing in our country. And upon hearing [about laws], I became very animated because it means that we aren't doing anything against the law, that we are within our rights [to make such demands]. And it made me very sad, for in our case they say that "any person who congregates is against the law, is a communist, is a bad person".'

Clandestine cemeteries

After several years of petitioning the courts and the Public Ministry by CONAVIGUA widows, GAM (Group of Mutual Support – the group of relatives of the disappeared) and Consejo de Comunidades Etnicas Runujel Junam por el Derecho de los Marginados y Oprimidos (CERJ: Council of Ethnic Communities 'We Are All Equal' for the Rights of the Marginalized and Oppressed) for an exhumation of what are estimated to be over 100 clandestine cemeteries in the highlands, the Assistant Human Rights Ombudsman, Lic. Acisclo Valladares, decided in June 1991, under a potential death penalty, to begin an exhumation of such a cemetery in Chichicastenango containing more than 100 cadavers of peasants assassinated by security forces. Although eight court-ordered exhumations had been carried out by the relatives themselves, this was the first public recognition of their petitions, thus rendering for the first time their *denuncias* legitimate (*El Gráfico* 21 June 1991: 4 and 8; *Siglo* XXI, 2 July 1991: 6).

As the GAM president, Nineth de García, stated:

'[Exhumations] are important for psychological and moral reasons. Not knowing where our loved ones are slowly eats away at us. It is better to know that they killed our relatives, as difficult as that is to accept, than to live with the pain of not knowing where they are.'
(quoted in Americas Watch 1991)

While military authorities in Guatemala have dismissed these graves as the burial grounds of guerrillas, they have at the same time threatened CONAVIGUA, GAM and CERJ members. For example, on three separate occasions, civil patrollers in the village of Pacoc, El Quiche, have tried to abduct CONAVIGUA member Juana Calachij for her role in arranging the exhumation of five bodies from a mass grave (including that of her husband) who were hacked to death by civil patrollers with machetes in May 1984. She continues to be under surveillance for her activities.[15]

Indigenism: a 500-year celebration of discovery of the Americas?

Another theme of the lectures that the CONAVIGUA directorate attend concerns the 'discovery' of America by Columbus, which the Guatemalan government celebrated with a large fiesta, calling it 'The Meeting of Two Worlds, Two Cultures'. As Juana, one CONAVIGUA directorate member, comments dryly:

> 'Celebrate what? To celebrate the spilling of the blood of our ancestors, the plundering of our lands, of our culture, of our customs, which we know concerns not only our ancestors, but continues into today. Thousands and thousands of brothers [*hermanos*] are still spilling their blood in seeking a more just life, in seeking a more human life. Thousands of *campesinos* are thrown off their land, knowing that these lands belonged to our ancestors: who knows where they are going to live? None of this has ended; we are still living this bleeding and plunder.
>   So we cannot call it a "A Meeting of Two Cultures"; we would prefer to see it as a beginning of the meeting of two cultures in which there is an end to the marginalization, exploitation, discrimination that we are living. That's what it means for us.'

As CONAVIGUA's July 1991 report states, 'We participate in the struggle of the last 500 years of Indigenous and Popular Resistance, as much on the National level as the International, in defence of our culture.'

Threats and the costs of defiance

CONAVIGUA began in the province of El Quiche, and now functions in six provinces. However, in Guatemala it can be dangerous as an indigenous woman to learn how to read, organize and apply the Constitution: 'They don't like us working together, acquiring knowledge about the law' (Lucia quoted in Frankel 1990).

In June 1990, Amnesty International expressed concern over the attempted abduction of fourteen individuals, the majority of whom were CONAVIGUA members. In March 1990, CONAVIGUA member María Mejía was shot in her home in Parraxtut, El Quiche, by two armed men who were recognized by her husband as military commissioners.[16] 'We are in danger of death simply for speaking out against the high cost of living and studying the Constitution and laws of our land' (ibid.).

As Juana says angrily:

'But not only that: we as women have been seen as objects, as only something to be used. Many women have been raped by the military authorities. They come here to our house [of CONAVIGUA in Guatemala City] to rape us. They say that that's why the house is here – we are women without husbands. So, they not only kill our husbands, but they come to rape us in our homes. All of this has been forgotten. No one wants to listen to the stories of our lives.'

Indeed, these women's literacy classes are a defiance on several levels of vulnerability: they have been raped by soldiers, military commissioners and *jefes* of civil patrols 'who know that the indigenous woman doesn't speak Spanish and thus cannot bring charges before a judge'.

However, the widows are further threatened if they report their rape (and pregnancy): 'They take advantage of us; they force us to cook for the soldiers at the garrisons and bases' (Lucia quoted in Frankel 1990).

COMPARISONS AND PARALLELS BETWEEN THE TWO GROUPS

Similar in their experiences as pre-literate, rural women, with some previous community- and religious-based organizing, the CoMadres and CONAVIGUA represent the new kind of popular forms of resistance by women arising in Central America over the last fifteen years which cross both class and ethnic boundaries.

Women in both organizations, given their experience with the brutalities of counterinsurgency, such as massacre and rape, recognize the need simultaneously to connect their demands to respect human rights with the need to demand a respect for women's rights and dignity, initiating a consciousness not only, it seems, among themselves as wives and mothers, but also among their daughters. This legacy of cross-generational consciousness apparent in both groups may be unique to motherist–daughterist movements in Latin America, although more comparisons need to be made.

These women rupture the image of the essentialist female victim or the passive witness to war by taking photos of cadavers for identification, by demanding the exhumation of their relatives in clandestine cemeteries, and by identifying by name those responsible for the killings.

Moreover, both the indigenous widows and the CoMadres threaten the cheap labour supply of army troops in Guatemala and El Salvador respectively by demanding an end to forced recruitment and the brainwashing of their sons.

Moreover, they both contest their status as sexualized objects by security forces, responding to the humiliating experience of torture and rape with dignity and anger. Indeed, strong mother–daughter bonds based on the rape experience are part of the legacy of this gendered consciousness.

Finally, as 'seekers of truth' as to what exactly happened to their relatives during '*la violencia*', the brutality of the security forces and the indifference of the civilian regimes has made them realize the need to protect themselves physically and culturally by becoming literate in both linguistic skills and the legal culture.

IMPLICATIONS FOR FEMINIST THEORY OF WOMEN'S ACTIONS: TAKING ISSUE WITH PAST FEMINIST DEBATES

If we assume that 'feminism' is a contested terrain, mediated through and contingent upon the shifting relations of class, race, political agency and political repression, how can we study change in women's consciousness without assuming that there is only one way in which women come to know the world and understand their experience? How can we do so without affixing a 'naturalized' agency or essentialized (and thus seemingly universalized) identity upon women? How do we study such change without 'imposing a direction, bending it to a new will, to force its participation in a different game' (Foucault quoted in Scott 1991: 796)? In sum, how do we study women's actions, leaving the definition of feminism open for discussion?

Both the relatives of the disappeared in El Salvador, the CoMadres, and the widows of political violence in Guatemala, CONAVIGUA, help us to understand how different kinds of experience sculpt various constituencies of feminist consciousness. Yet what are the feminist categories we have been provided with by which to understand such actions by women?

Political motherhood: how political can mothers be?

In dealing with the issue of 'motherist movements', or how political can mothers be, some feminist analysts have tended to argue in two directions, yet have remained within the same cultural essentialist paradigm.

The thesis of 'maternalist women' draws upon an anti-militarist model of mothering within 'politically stable' states. Maternalist feminists have focused upon the cultural essentialism of all women-as-caring-mothers, able to nurture and be ethically responsible (Ruddick 1984 and 1990). Jean Bethke Elshtain, building upon Hannah Arendt's concept of birth as natality, argues a gendered essentialism that perceives a betrayal of trust in a violent death (Elshtain 1982). She refers to a maternalist experience that 'on an even more basic level, taps a deeply buried human identity' (ibid: 56). In another essay, she argues for feminists to break free from the Kantian ghost of absolutist binary constructions, contrasting masculinism and violence with feminism and harmonious order (Elshtain and Tobias 1990: 265). Yet Elshtain offers little insight as to how one goes about breaking out of this Western way of seeing.

Similarly, while Ruddick acknowledges that maternal thinking (or what she terms 'the rationality of caring') has been denigrated and devalued by Reason, she goes on to argue that 'the political effectiveness of care depends upon a wider feminist and labor politics that emboldens caretakers – especially women' (Ruddick 1990: 249). However, we are not informed what the nature of this 'wider feminist . . . politics' is and who gets to define it. In other words, women's 'social actions centered around caring' can indeed be varied and account for 'difference' (ibid.: 238), yet, once these 'caring' groups 'go beyond' the particular into a 'broader' public arena, 'feminism' is presented as an essentialist ethos. 'Feminism' for both Sara Ruddick and Jean Elshtain appears to be 'frozen' in time and space. For all its claims to include 'difference', the maternalist feminist argument, in the end, ignores the contextualization of mothering and thus misses the opportunity to go beyond its own cultural assumptions to deal with the nature of potential actions by politicized mothers within repressive circumstances.

The other feminist argument of 'complicitous women' tends to focus upon how a repressive state, such as Nazi Germany, comes to appropriate 'motherhood' for its own purposes. Claudia Koonz, for example, argues that 'mothers and wives made a vital contribution to Nazi power by preserving the illusion of love in an environment of hatred'. They were the 'significant other' of mass murderers. That is, mothers acquire horrific political responsibility by loving.

While Koonz has not succumbed to the cultural essentialist feminism that makes women especially morally virtuous, 'she succumbs', writes Victoria de Grazia, 'to another kind of exceptionalism that makes the Holocaust historically unique and subjects anybody connected with it to special moral imperatives' (de Grazia 1987: 508–10). By implication, does this thesis assume that women are especially irresponsible if they do not oppose war or political violence? Are we to hold all women whose husbands flew in Vietnam or the Gulf War indirectly responsible for the carnage of those wars? While this position allows us to analyse repressive states' expropriation of 'motherhood', it unfortunately assumes women's agency-by-complicity if women do not openly and visibly oppose war.

These arguments are, frankly, not germane in helping us understand women's actions in Central America. Surely, one can speak of 'motherist/widow' movements without necessarily falling into a maternalist feminist trap – i.e. one that argues that women, by virtue of their potential to be mothers, are especially ethical and responsible towards life. Nor need one argue that mothers' and wives' complicity in 'allowing' their sons to be forcibly conscripted into the army in Central America are particularly irresponsible.

Feminist or pragmatist: a false duality

Other feminist analyses of women's political activism have tended to assume exclusive, binary categories. 'Female consciousness' and 'feminist consciousness' (Kaplan 1982), as well as 'strategic interests' and 'practical interests' (Molyneux 1985)[17] were important contributions to feminist vocabularies in the mid-1980s in attempts to decipher the significance of particular kinds of consciousness that exist among politically active women. While adding to our understanding of why and how women protest, women's actions have tended to be separated into those who act out of 'broader', feminist 'strategic' concerns and those who act out of more 'female, pragmatic', social and economic concerns. Unfortunately, they are too often dichotomized as feminist or female/non-feminist, theorist/strategist or pragmatist/social.

If the intention is to understand if, when and how women's consciousness changes, and what the nature of that consciousness is, and whether it is 'feminist' or not, then I would like to suggest that such dichotomies tend to force a hierarchically structured set of expectations, with the 'final' ascendence of Feminism (with a capital F and with Western assumptions as to the nature of that Feminism). Moreover, it tends to assume exclusionary interests: that one cannot move from one category into another, or be at once 'pragmatic' and 'strategic', 'female' and 'feminist'.

The need to move beyond the binary assumptions

One of the political borders that feminist analysts have yet to cross, within the undulating and often underground topography of women's political consciousness, is in the understanding of not only how *women who may begin as pragmatists become strategists*, but also how women who begin with the pragmatic and strategic struggle for human rights (for example, the right to life centred around family survival, exclusive of the conscious demands for the rights of women) learn of the need to connect these struggles to the pragmatic and strategic struggle of women's rights; that is, how to make politics at once political and personal, at once strategic and pragmatic – and then coming to realize that they may have been 'feminists' all along!

For, given these binary categories, how are we to understand women who grow beyond one ideological world and seek out another? That is, women who have *multiple* 'strategic' and 'pragmatic' interests that change over time and that they themselves do not deem separate? For example, although Christian humanism and the legal community of the human rights movement are central for the relatives and widows of the disappeared in organizing around their grief, these women reach certain junctures in their analysis about political violence where teachings of Christian humanism and the legalistic approaches of the human rights movement do not address their specific needs as women[18] – such as the experience of rape. This is not to say that the women necessarily reject such teachings or approaches; they supplement them with their gendered experience of political violence.

Moreover, within the midst of *la violencia*, they have established daycare centres, health clinics, rape discussion groups and battered-wife support groups in El Salvador, and a woman's house for widows in Guatemala. Are these changed meanings and constructions to be dismissed as 'pragmatic, female interests'? Can we not speak of womanist feminisms (with a small 'f') which often (but not necessarily) bridge these two interests?

Finally, the individualistic vision of liberal feminism cannot categorize what could be called the CoMadres's and CONAVIGUA widows' strong sense of collective citizenry which entails the obligation to pass on the collective memory of such impunity to their daughters and sons so that the past will not be forgotten – and allowed to be repeated. This added sense of responsibility for their future children and grandchildren, it could be said, reflects a kind of *collective citizenry of political motherhood* – and potentially fatherhood in their future vision – in which they believe that there can be no national security without social justice. Thus, in this non-binary vision, Family and Justice are collapsed into one moral domain.

CONCLUSIONS

The CoMadres and the CONAVIGUA widows allow us to see how in the seeking of truth and justice for their families, women can gain a gendered consciousness of political woman/motherhood and a responsibility of collective citizenry that is being passed on to their daughters and sons. They may also allow us to understand how the transformative process of gaining a gendered consciousness is dialectic, not binary – that is, dependent upon both what the women bring to the struggle, as well as how the state constructs them and their actions externally (see Radcliffe, this volume). For example, within the process of struggling as mothers, as former wives, they themselves have been sexualized, brutalized by the security forces. The need to protect themselves as women forces them to develop first an analysis of how and why they are abused by the repressive state. The horrific experience of rape and torture is turned into organizing around investigating, as CoMadre Alicia says, 'the bacteria that is introduced into our bodies [by security forces]'. It is analysed in terms of 'if they can't silence us one way, then they try to humiliate us as women sexually'. This analysis quickly extends to a larger analysis of how women are so often abused by their employers (especially the *finqueros* on the large plantations where the widows have worked), and by their husbands (in cases of battering of the CoMadres). What was originally a 'pragmatic' necessity to protect specific mothers and widows from sexual abuse, became a 'strategic' necessity to protect the gender 'women' from such generalized and accepted violence. What were initially seen by these mothers/widows as individual acts of political rape became perceived as part of a culture of gender abuse.

The construction of consciousness

We know that a change in a person's consciousness assumes that different meanings and constructions are in conflict. We either readjust our vision to incorporate them and thus resolve the conflict, or we come to learn from experience that no matter how often we readjust our vision, the conflict cannot be resolved unless we eventually ask who is in control of these meanings and constructions – and ask what is, for many, the ultimate feminist question, 'who is claiming the truth?' (Hawkesworth 1989). If the old patterns and identities don't 'work' for us anymore in providing us with 'the familiar', we ask ourselves, why don't they?

The sense of the 'natural' role of mothers as apolitical and submissive remained constant among the women spoken to here until it conflicted with the loss of their children and families to massacre and disappearance. At this point, the security forces sexualized and abused these women while at the same time denying that such actions ever occurred or that any such person existed. These women's ways of knowing the world, their vision of self-identity, then collapsed, and for a while they were at a loss as to what to do. As one CONAVIGUA member states, 'The first thing we had to conquer was our own fear.'

These women overcame their fear and began to construct an alternative script to replace the one worn out by abuse. They began to ask why.[19] By refusing to accept the boundaries of the state as to what they were not allowed to know, in the process they both exposed the ethical vacuum of the state while gaining a gendered consciousness. Each time, in pushing beyond this denial and censorship of their known experience, they learned more about these 'claimers of truth', and they began to 'think about what we were doing as women'. Their sense of 'knowing', of learning from each other's experience, which was in conflict with 'the truth', was continually reconstituted, especially as patterns of violence against them began to emerge. The question of 'What is the truth about our relatives?' slipped into 'If they deny us the truth about our relatives, then what else have they lied to us about?' and 'If they say we are mothers who should be respected, and yet treat us and our daughters with rape and torture, who are these men who sexualize us, soil us and degrade us?' In this process of questioning first 'the truth', and then the 'claimers of that truth', class and ethnicity gained and lost their centrality to gender, sometimes returning in the form of gendered ethnicity or gendered class, but increasingly seen through the optic of gender.

Although, as Joan Scott (1991) warns, not everyone learns the same lesson from this 'conflict of meaning', nor do we learn it at the same time or in the same way; it seems to be the case that state repression has created a particular kind of gendered consciousness in which claimers of truth are challenged by a significant number of human rights groups led by women in many countries in Latin America.[20]

One could argue that these women, in seeking 'to know who claims the truth', come to ask 'feminist' questions of power relations that stretch far beyond the immediate search for relatives.[21] Indeed, claims of truth are at the basis of much feminist criticism (see Hawkesworth 1989). In rethinking and challenging the constructions of 'mothering' and 'fathering', 'femininity' and 'masculinity', 'security' and 'justice', the mothers and widows are making connections between women's rights and human rights, between family/private and justice/public.

Yet, if Western feminism as we know it arose from particular US and European political circumstances, can it be translated 'frozen' into an understanding of how women under political circumstances particular to Guatemala and El Salvador make sense of their gendered worlds? Should we assume that Western feminism has a priority of 'knowing'? Do Western feminists constitute an elite with a superior understanding of theories of 'truth'? Does Feminism in this regard act as a kind of Leninism in disguise asking 'What Is To Be Done?' rather than, as Hannah Arendt suggested, 'Think What We Are Doing?' (Elshtain 1990: 255).

Rather, shouldn't we want to know more about how women make sense of their conflicts, how they think about what they are doing and who they see as having the power and ability of making sense of these actions, and how they come to gain a positive self-image? Shouldn't we

63

3

ECOLOGIA: WOMEN, ENVIRONMENT AND POLITICS IN VENEZUELA[1]

María-Pilar García Guadilla

INTRODUCTION

In spite of the evidence that women in Latin America have been active in most social movements, sometimes even more active than men, they have been excluded from leadership roles and political power associated with such movements. The literature has stressed the fact that women are linked to political power in weaker ways than men (Kaplan 1982; Bartra *et al.* 1983; Logan 1984, 1988, 1990; Shiva 1987, 1989; Molyneux 1985; Leacock and Safa 1986; Lozano Pardinas and Padilla Tieste 1988; Bennet 1989; Alvarez 1989; Rao 1989, 1991). Lack of effective power for women has been associated with a combination of two main factors: women's reduced access to political power, and the domestic nature of women's demands, that limit the ways that these can be transformed into political facts.

Throughout this chapter, I will contend that many demands, by the sheer fact of having been appropriated by women, are associated with the domestic domain, whether or not they belong objectively to this domain. When appropriated by men, the same demands are reclassified as political.[2] However, in the case of the ecological or environmental movement, the main factor explaining whether demands become a political fact is not gender, but the strategy displayed and the use of the mass media. More than gender, the conversion or not of these demands into 'new' political problems seems to depend upon: structural characteristics, strategies and leadership's style of organization, strength of the social relationships within the group; social class composition and access to financial resources; ideological homogeneity or heterogeneity of the members involved; potential for articulation with other organizations and, above all, access to and use of the mass media.

The ecological or environmental organizations in general can be traced back to the 1930s, and more directly to the 1960s (García 1986).[3] They range from pioneering scientific conservationist societies which reached their zenith toward the end of the 1980s, to the more recent, so-called 'symbolic' cultural organizations which are reconstituting the 'environment' as a new political fact (as defined by Melucci 1985; Uribe and Lander 1991; García Guadilla 1991a). If we define the environmental organizations of the last thirty years in a broad sense as organizations that pose physico-natural or socio-environmental demands, and if we group them according to sex composition, socio-economic, ideological and cultural characteristics, organizational structure and other variables, we find today at least six different types (García Guadilla 1991a, 1991b, 1991c). There are scientific-conservationist societies; ecological communities; environmental defence juntas; political ideological organizations; symbolic-cultural organizations; neighbourhood associations and women's environmental groups. The last two types have dual demands: the installation or improvement of urban services (neighbourhood organizations), or the achievement of greater sex equality and justice (women's environmental groups).[4]

With the exception of the symbolic-cultural and the women's environmental organizations, in all the others (particularly in the political-ideological) men tend to be leaders. In neighbourhood associations although women comprise more than half the membership, over half the leaders are men, while in the political-ideological organizations males tend to predominate among both leadership and the membership.[5] I contend that women differ from men as potential political actors (Logan 1990; Rao 1991) and that they have different political impacts on the government and on public opinion according to their strategies, identities and capacity to transform their demands into new political facts.

There are at least three approaches from which social organizations in Venezuela have incorporated environmental demands: the ecological, the gendered and the urban.

To assess the consequences of the increasing 'politicization and environmentalization' of urban demands, a comparison will be made between low-income squatter neighbourhoods and middle and upper-middle income groups. These two processes seem to work in a contradictory way. First, the 'politicization' of urban demands is the result of the allocation of power to neighbourhood organizations through new legislation which has attracted more men into leadership roles, particularly in the middle- and upper-middle-income neighbourhoods.[6] Second, the 'environmentalization' of urban problems in Venezuela coincided with the economic crisis of the 1980s and with the growth of poverty and lack of basic services, water shortages, falling sanitary and health conditions, in common with main environmental problems elsewhere in the Third World (Sunkel 1983). Given its identification with poverty, the environmentalization of urban demands is occurring more among the low-income neighbourhood associations and among women (Kaplan 1982;

Logan 1990; Rao 1991) making it difficult to differentiate between urban and ecological demands. Although the 'environmentalization' of urban issues is bringing more women into neighbourhood organizations they do not have any formal power. Men, who have been and still are the traditional leaders, are interested increasingly in positions of leadership (Logan 1990). Finally, we will compare men and women who participate in neighbourhood associations to assess their strategies and identify the existence or not of 'invisible networks' (García Guadilla 1992).

I include five types of organizations: a) scientific-conservationist societies; b) ecological communities; c) environmental defence juntas; d) political-ideological organizations and e) symbolic-cultural organizations. To study the influence that the gender composition of the leadership has in creating the environment as a 'new' political fact I will compare the political-ideological, and symbolic-cultural groups that are led by women.[7] Ecological organizations led by women tend to be more successful in opening new spaces of political significance, that is in transforming the ecological into a new political fact (Melucci 1985; García Guadilla 1991a). This success is due to the structure and strategies displayed by the organization that include the use of mass media.

In terms of gender, I evaluate the effects of the dichotomy of gender/environment on the creation of the ecological as a new political fact by comparing women's organizations with women's environmental groups. In spite of the potential gender has to create new political facts, when both gender and the 'environment' have combined as a joint demand this potential has decreased.[8] In other words, in women's environmental groups the creation of new political facts has diminished.

In order to explain the potential each approach has to influence political culture and to define the 'environment' as a new political fact, I will evaluate the effect gender has on the identity, organizational structure, social relationships, strategies and other ideological and socio-economic characteristics in the social movements. Finally, I address, briefly, the way in which Venezuelan environmental organizations have evolved throughout the last three decades in their strategies, practices and identities, to be more efficient politically in the changing socio-political and economic conditions.

VENEZUELAN POLITICS: CRISIS AND SOCIAL MOVEMENTS

Thanks to the significant revenues generated by oil exports, during the 1960–80 period Venezuela was one of the few Latin American countries able to maintain relatively high rates of economic growth. The economic bonanza, together with external loans, helped to finance the government's modernization and industrialization policies based on the strategy of industrial and mining poles of development (Friedmann 1961; García Guadilla 1986, 1987). Venezuela is also one of the few Latin American countries that has maintained a 'formal' democratic system during the last thirty years. The establishment of democracy in 1958 failed to open up a political space for civil society, however, because the democratic value of participation was sacrificed in the 1961 Constitution on behalf of the stability of the political system (García Guadilla 1989). The representative democracy institutionalized in 1961 had strong paternalist, populist and clientelist features (Rey 1987).

Nevertheless, the political and economic models introduced in Venezuela since the end of the 1950s have created a great heterogeneity of social sectors with divergent interests (García Guadilla 1986, 1987, 1990). The autonomous growth of civil society, however, has been impeded by the cooperation and institutionalization of social conflicts through socio-political pacts among the main power blocs: the armed forces, the Church, the parties, the private sector and the unions (García Guadilla 1987; Gomez 1991). This has resulted in the structural weakness of all social organizations excluded from such pacts (de la Cruz 1988; García Guadilla 1990; Gomez 1991).[9]

The deepening of the economic crisis in recent years (in gestation since the end of the 1970s with the sharp decline in the price of oil) pressed the Venezuelan government in 1989, at the request of the International Monetary Fund, to adopt drastic macroeconomic adjustments which resulted in a greatly deteriorated quality of life for most people. As a consequence of these measures and the lack of political institutional channels to express popular discontent and frustration, strong social protests erupted in the same year (García Guadilla, 1990, 1991a), which placed the Venezuelan political system at a crossroads between democracy and authoritarianism.

One of the most important consequences of the economic and political crisis was the restructuring and redefinition of power relations. As part of this process, the previous socio-political pacts fissured, an increasing mistrust of the political parties arose in civil society, and parties lost their ability to stand for the transformation of representative democracy into participatory democracy, emphasizing local spheres of participation. A new, increasingly complex political scenario arose due to the heterogeneity of interests, making consensus difficult and negotiation within the power bloc necessary. New forms of mobilization and protest appeared including the violent social outbreak of 27 February 1989, through which popular protest discovered a new potential for mobilization in the face of the government's institutional incapacity to channel their demands (García Guadilla 1989, 1991a, 1991b; CENDES 1989). The environmental and political nature of these popular demands are related to the damaging effects of the crisis on the quality of life and on the satisfaction of basic needs that are the most significant environmental and political issues throughout the Third World (Slater 1985). The fulfilment of basic needs in Venezuela, as elsewhere in Latin America, is conditioned by the need for greater democratization of the political system (Torres and Arenas 1986; García Guadilla 1988a).

To achieve these political-environmental demands, new forms of articulation have emerged among popular organizations, neighbourhood associations, grassroots Christian groups, women and environmentalists. The mistrust of the old socio-political model, the rejection of the political parties, and a deepening economic crisis have gradually delegitimized the 'political' space as a privileged place to address national problems. These very factors have displaced the party system as the only realm in which to create shared meanings about a desirable society. As a response to these challenges, in 1984 the government created the Comisión para la Reforma del Estado (COPRE: Commission for Reform of the State).[10] COPRE's objectives were 'to adopt measures to assure the establishment of a modern, essentially democratic, and sufficient State, in which the postulates of the Constitution may enjoy full observance and the citizens' participation may constitute an effective element in decision-making processes of public authority'.[11] This commission has been responsible for directing political, economic and administrative decentralization, considered the most suitable means for democratizing and modernizing the Venezuelan state. Even though the commission's objectives incorporated civil society's aspirations, political pressures slowed down its achievements.

The organization of women in Venezuela may be traced back to the 1940s when they demanded and got the right to vote in 1942. In the late 1950s, women joined the underground forces against the dictatorship and for the struggle towards democracy (their slogan was '*Nosotras también nos jugamos la vida*' (We [women] will also risk life)). At the beginning of the 1980s, there were also important mobilizations of women in favour of changing the unjust Código Civil (civil code) to achieve greater equality supported in the constitution. Those changes were achieved in 1982 with the publication of a new civil code.

During the last thirty years of democracy, a variety of women's groups have emerged. Women have organized through labour unions, in favour of human rights and peace movements, against violence; through popular action groups to improve their life conditions in the squatter settlements, to discuss feminism; by forming lawyer's federations, journalists' and physicians' associations, parliamentary-political groups, and subgroups in political parties, among other ways.

One of the movements with the highest profile was the feminist movement, arising in the second half of the 1970s. It encompasses primarily middle-class intellectuals with postgraduate studies in the USA,

France and Italy, and leftist women who had renounced leftist parties where feminism was considered 'a capitalist deviation'. Many were from overseas, living in Venezuela and working as filmmakers, photographers and/or in the cultural industry, and some brought professional backgrounds in the social sciences, particularly sociology. The influence of the French liberation movement of the 1960s and 1970s was very strong since many of the founders had participated in this (Grupo Feminista Persona 1978–80; Grupo Feminista Miércoles 1978–91; Encuentros Feministas Nacionales 1979, 1981, 1985, 1987, 1989; Espina and Patiño 1984; Merola 1985). Women from the two most important feminist groups, Grupo Miércoles and Persona, were related by strong bonds of friendship, as well as professional and ideological concerns. As the groups got larger and more heterogeneous, these social relationships changed in form. However, in the case of the Grupo Miércoles that survives to this date, and friendships have played an important role in the maintenance of the group.[12]

In 1985 and despite wide divergence in origin, ideology and socio-economic composition, most women's organizations came together and constituted the Coordinadora de Organizaciones no Gubernamentales de Mujeres (CONG: Coordinator of Women's NGOs).[13] Today, this *Coordinadora* includes more than forty organizations of women whose interests range from the religious to the professional (CONG 1991). The main objective of CONG is 'to associate all women's organizations, who worked towards ending discrimination in the legal, economic, social, political and cultural areas as well as to coordinate activities to defend women and to help them to achieve full participation in the country's active life under conditions of equality' (CONG 1991). CONG, in order to deal with the great heterogeneity of organizations which could lead to potential conflicts, established that each organization was autonomous and independent when pursuing its own objectives. To achieve the collective objectives, a horizontal process of making decisions, whereby all decisions were taken by consensus, was designed. Although in yearly assemblies, such as the one held in 1989, there were substantial differences, the friendly social environment, cooperative 'style' and development of strong personal interrelations that characterized those meetings helped to achieve full consensus.

The importance of the *Coordinadora* has been to provide a juridical and institutional framework for women's groups with the legal endorsement of CONG as a non-governmental organization in 1988; to share and distribute relevant information quickly to all the groups involved; to bring together women with different interests, socio-economic background and ideologies, and to negotiate with the government from a position of power that derives from being a collective body grouping the largest number of women's organizations.

The first three goals have been successfully achieved by the *Coordinadora*. However, the fourth seems more difficult to achieve given the inherent contradiction it poses. While it seems to be true that by sheer quantity the aggregation of new groups brings more political power to the *Coordinadora*, it could also weaken it. This would be the case if the groups were too heterogeneous in their interests making consensus difficult, or if they raised demands which might create competition for scarce funds, power or recognition among the different groups involved in CONG. If left unresolved, these internal conflicts could reduce the potential of CONG to transform the political culture and to permeate it with women's demands.

ECOLOGICAL, FEMALE AND URBAN APPROACHES TO THE ENVIRONMENT

Just as in other Latin American countries, the origin and characteristics of the more than 100 environmental organizations that exist in Venezuela (García Guadilla 1988a) are related to economic, political and social factors. In Venezuela, the following factors stand out: the economic strategy of development based on government sponsored large-scale industrial-mining projects; the environmental impact of this strategy; the incipient level of democratization achieved by the political system; the low degree of complexity of civil society *vis-à-vis* the high level of institutionalized and juridical development concerning environmental issues; and the growing degree of popular awareness about the environmental question (García Guadilla 1991a).

Similar to Brazil and Mexico (Slater 1985; Viola 1987; Quadri 1990; Puig 1990; Gerez-Fernandez 1990), the environmental organizations of Venezuela are highly heterogeneous, both intra- and inter-organizationally, from the socio-economic, ideological and cultural points of view (García Guadilla 1991b, 1991c). On the other hand the formal structure of Venezuelan environmental organizations varies according to the organizational and 'participatory' opportunities offered by the political system, the organization's ideological background, and the gender composition of its leadership. Organizations with women in leadership roles tend to be more flexible, horizontal and less 'personalist' than those where men are the leaders. Similarly, gender composition at the leadership level seems to be correlated with the adoption of strategies of negotiation or confrontation. Symbolic-cultural organizations, such as AMIGRANSA where women leaders predominate, tend to adopt negotiating strategies, propose alternatives, and use institutional channels more frequently than those organizations where men predominate. The latter organizations are orientated mainly toward confrontation and they maintain a more defensive character.

71

POLITICAL-IDEOLOGICAL ORGANIZATIONS

Political-ideological organizations emerged at the end of the 1970s and the members were primarily either young men who had abandoned the 'guerrillas' and taken advantage of the pacification process or university men from the leftist parties (USB-ILDIS 1987; Viola 1987; Slater 1985; Quadri 1990).[14] Because of these ideological origins, men of these organizations tended to be more orthodox, 'personalist', rigid and possibly *machistas*. This could be connected to the fact that during the 1970s the leftist parties (Movimiento Hacia el Socialismo (MAS: Movement To Socialism); Partido Comunista (PC) and Liga Socialista) considered the environmental and feminist demands as petit-bourgeois deviations; consequently, they barred both demands from party politics. However, some members from the leftist parties, who were predominantly men, decided to form environmental organizations to channel those specific demands while remaining as members of the parties. They did not question the *machista* practices already existing within the parties from which they came; therefore, they tended to reproduce them inside the environmental organizations. Women, on the other hand, were very dissatisfied with the prevalent *machismo* inside the leftist parties and considered it impossible to reconcile this with their general approach to politics; therefore, they broke out of the parties and formed the feminist movements (Donda 1978).

Political-ideological organizations share an eco-socialist view which criticizes the style of capitalist development in Venezuela as the cause of the gradual worsening of living conditions, the marginalization and impoverishment of large sectors of the population, and the growing deterioration of the physical and natural environment. The solutions they suggest are thus aimed at transforming the economic and socio-political model into a more socially egalitarian and technologically rational one, as embodied in the notion of 'eco-development' (García Guadilla 1991b, 1991c; Grupo Ambientalista Eco XXI 1987, 1988). Seemingly, they are not aware of the apparent contradiction which exists between eco-development and the structural transformation of dominant development models (Leff 1991). Some examples of this type of organization are the Grupo de Ingenieria de Arborización (GIDA; Group of Forestation Engineering and Habitat). Usually, they have assumed a defensive position and they have been against mega-projects such as the exploitation of the Orinoco Oil Belt, and the petrochemical, coal and gas industries. Since the 1960s, those mega-projects have been, and still are, the core of Venezuela's economic model. Although the organizations mobilized people against those mega-projects, they were unsuccessful in achieving their goal of transforming the economic model (García Guadilla 1987, 1990). On the other hand, their access to the mass media was rather restricted and their demands did not transcend the environmental lobbying generated by the Ministerio del Ambiente y de los Recursos Naturales Renovables (MARNR: Ministry of Environment), the state-oil company of Petroleos de Venezuela (PDVSA) and the environmental groups.

Their major success was the constitution of a federation, Federación de Organisaciones de Juntas Ambientalistas, (FORJA: Federation of Environmentalist Organizations) that grouped more than 100 different organizations. In spite of the democratic-horizontal structure that was devised for the federation, the problems of personalism, authoritarianism, *machismo*, ideological rigidity, and competition for leadership already present in the individual organizations were reproduced themselves in the federation, making it weak (USB-ILDIS 1987). Many of the leaders of the member organizations belong to political parties of the left which during the 1970s rejected environmental demands. Conversely, during the second half of the 1980s they recognized their political importance, in view of the international relevance the topic has acquired and pending approval of the Ley Penal del Ambiente (Penal Law on the Environment) in Venezuela that would reinforce the already existing strong environmental legislation.[15]

Another success of FORJA has been the entry, in the role of technical advisers, of some members of the political-ideological organizations into the Comisión de Ordenación y Ambiente del Congreso de la República (Congressional Commission on Environment). As part of the legislative body, this Environmental Commission plays a major role in relations with political parties and acts as a mechanism against the Executive. This 'double militancy', in environmental movements and in political parties, of some members of the political-ideological organizations who work as technical advisers for the parties represented in the Commission, maximizes the effectiveness of such a relation. However, this double militancy makes some of the individual organizations of FORJA, particularly the ones that could be defined as 'political-ideological', more vulnerable to cooptation by political parties. It is striking that all members of the political parties that comprise this Commission, as well as the technical advisers from the environmental organizations, are men. Moreover, all the technical advisers belong to the leftist political party Movimiento hacia el Socialismo (MAS: Movement to Socialism) and to political-ideological environmental organizations.

Political-ideological organizations are more interested than other ecological organizations in pursuing 'traditional' political power, through members' participation in political parties and through the Environmental Commission. Paradoxically, their demands for the transformation of the model of development are considered too radical to be channelled through political parties of the left: therefore, they are channelled through the ecological organizations.[16] Since they usually penetrate the institutional system of party politics through the Environmental Commission, they do not create a new political space for environmental demands. This traditional style of doing politics, which as we will see differs sharply from the one developed by the symbolic-cultural organizations, has proved to be unsuccessful in achieving their demands.

SYMBOLIC-CULTURAL ORGANIZATIONS

In 1985, with the lack of answers to the national environmental problems, the loss of credibility of political parties and the government as well, and the weakening of politico-ideological organizations, new symbolic-cultural organizations emerged.

They are referred to as 'symbolic-cultural' because they alert society to the existence of problems whose solution requires decoding the dominant models and searching for alternative meanings and orientations for social action in the cultural sphere (Melucci 1985). Furthermore, these organizations fall within what Uribe and Lander (1991) call 'new domains of the political', evident in the emergence of forms of social action independent from both the political parties and the state, and which revolve around new themes. As in the case of the environment, these forms address problems:

> that can be acted upon or influenced without going through parties or their mediation, and without the need to validate proposals against the backdrop of a larger social project and its associated strategy.
>
> (Uribe and Lander 1991: 77)

This type of organization is best exemplified by AMIGRANSA, a non-profit making Civil Association founded in 1985 by only five members, all of them women, who form the executive committee and design the group's strategies. Most of AMIGRANSA's actions have focused, successfully, on the defence of Canaima National Park, the world's fifth largest (AMIGRANSA 1989, 1991). In contrast to the political-ideological organizations. AMIGRANSA does not question structurally the prevalent model of development; consequently, its demands can be legitimized within today's economic and political system. On the other hand, in all confrontations with the government over the environment, AMIGRANSA has proposed alternatives, most of which have been accepted.

Why, then, has AMIGRANSA been successful in contrast with the political-ideological organizations? The answers lie in two main areas: gender-associated strategies and the effective use of the media. AMIGRANSA's success lies in its ability to access and use information

74

and symbolic systems, particularly the media, as the main mechanism through which the 'environment' is constructed, produced and consumed as a new cultural and political fact. This is a new style of environmental politics since it relies not on traditional politics, like the political-ideological organizations, but on the transformation of political culture through the mass media and the generation of the 'environment' as a new political phenomenon. AMIGRANSA's composition by class and perhaps gender and occupation (middle- and upper-middle-class professional women), its greater availability of economic resources and time, its access to sources of financing and high-level government officials, and its connections with the mass media, have all contributed to the widening of AMIGRANSA's communicative space.

In contrast to the political-ideological organizations, the leaders of AMIGRANSA run less risk of being coopted by political parties because they do not belong to any of them. This gives them greater autonomy from party politics and more flexibility to deal with different political spheres, be it the government's institutions (MARNR), the legislative branch where the political parties interact (Environmental Commission) or the judiciary bodies. Another factor that has contributed indirectly to AMIGRANSA's success and might be related to gender composition is the pre-existing primary interpersonal relationships based on friendship of the group members. These social bonds provided the support for the constitution of a 'secondary' or 'interest' ecological group.[17] Because of the previous social and personal bonds, AMIGRANSA tends to be more homogeneous than other organizations in terms of age, occupation, ideologies, socio-economic composition and other personal factors. This coalescence has reinforced the organization's stability, facilitated internal consensus in decision making, and allowed a fruitful articulation with other ecological organizations around their demands, demonstrated in two significant actions: the Rally Transamazónico and the Spilberg-Tepuyes problems.[18]

However, due to the political and economic crisis, it is becoming more difficult for AMIGRANSA to gather support for their agenda. The ideological differences between the political-ideological, who favour socio-political objectives, and the symbolic-cultural, who favour physico-natural objectives, are getting wider with the crisis.

WOMEN'S ENVIRONMENTALIST ORGANIZATIONS

'Today, all women's groups in Venezuela are environmentalist, regardless of whether they know what the environment means'.[19] This comment, made by one of the founders of the *Coordinadora* CONG reflects the growing interest of women's organizations in environmental problems. It also expresses the growing identification of women's problems with environmental or ecological issues. Only in 1990–1 did Venezuelan women become interested in environmental problems from a gender perspective, in terms of the identification of women with nature.

It was not until 1990 that the Encuentros Latinoamericanos de Mujeres (Latin American Women's Meetings) fully incorporated the analysis of environmental problems from women's perspective.[20] Preparation for the United Nations Conference on Environment and Development, UNCED, held in Brazil in 1992, might have contributed to creating awareness of environmental problems among the non-governmental organizations (NGOs) throughout the world. In addition, international agencies such as the United Nations, the World Bank, the Inter-American Bank of Development, and the European Community, among others, have increasingly concentrated their attention on the subject of women and the environment. As a consequence of these trends, new organizations of women have emerged in Venezuela during the last quinquenium which focus on the environment, including Los Círculos Femeninos Populares, (Popular Women's Circles) created in 1988, Red de Mujeres Todas Juntas from the Centro de Formación al Servicio de la Acción Popular (CESAP: Centre for Training for Popular Action), Grupo de Estudio de Mujer y Ambiente (GEMA: Study Group on Women and Environment) founded in 1989, the Comisión Femenina Asesora a la Presidencia de la República (COFEAPRE: Female Presidential Advisory Commission) created in 1990, and the Asociación Venezolana Mujer y Ambiente (AMAVEN: Venezuelan Association of Women and Environment) founded in 1991.

Círculos Femeninos Populares y Red de Mujeres Todas Juntas

Los Círculos Femeninos and Red de Mujeres Todas Juntas are part of CESAP, one of the largest and better organized NGOs of Venezuela, created in the 1960s with the purpose of helping people to organize in the popular sectors or squatter settlements in order to raise their living standards. Círculos and Red are led and organized by women, and are part of CONG. They also work with women and their aims are closely linked to the improvement of the quality of life in the squatter settlements: health, contamination, lack of recreational facilities and parks, housing, employment, the high cost of living and education.[21] However, despite their aims and gender composition, they do not define themselves as environmentalists or as feminists.

The leadership of the larger organization CESAP is predominantly male and does not welcome the division of the poor or 'popular' population by gender. As was the case in leftist political parties and the environmentalist groups, the male leaders of CESAP believe that the incorporation of specific female, or gender, demands could disturb the

76

general purposes of the organization. For this reason, women from Círculos and Red have been afraid of being labelled 'feminists' since the beginning of their participation in CONG.[22] However, in the course of working together with feminist groups in CONG, they have developed a solidarity with women's problems and agendas and have implicitly started to question male leadership, the gender division of labour, and practices existing in CESAP. A source of conflict is thus created since the leadership of CESAP tends to minimize women's problems in the popular sectors while the process of conscientization within Círculos and Red tend to stimulate this issue. The same ambivalent attitude has been maintained towards the environment: the identification of poverty with environmental problems is relatively recent.

As in the case of AMIGRANSA, women from Círculos and Red tend to share strong bonds of solidarity and friendship. They have been very effective in proposing new alternatives to deal with the economic crisis in addition to defensive strategies against macro-economic structural adjustment policies (cf. Moser in this volume). Unlike AMIGRANSA, the issues raised by Círculos and Red do not have symbolic value beyond women's interests because they are identified with the domestic domain. Therefore, the transformation of their everyday problems into political issues depends upon the incorporation of such problems into the demands of the larger organization, CESAP. In these circumstances, a transformation has to come from the male leadership but, given their ambivalent attitude towards women's specific problems, it is unlikely that they will carry forward such a transformation. For financial support, the women started relying on the umbrella organization of CESAP. However, the combination of 'popular' and gender demands and their registration in CONG have been very effective in securing finance via international organizations such as the Inter-American Foundation.[23]

GEMA

Grupo de Estudios Mujer y Ambiente (GEMA), like AMIGRANSA, is a 'primary' social group of professional women founded by a few close friends in 1989. It has a flexible structure which facilitates consensus and is also a member of CONG. Recently, one of the members has been selected to be part of the Committee of AMAVEN. The main projects of GEMA are related to health and the environment in two squatter settlements in Caracas, and the denunciation of the precarious situation of women in the mining region of Guayana. The scale of their projects is small because the group has only three or four active members; it has to rely on the members' spare-time and funding, and does not have the backing of a strong organization such as Círculos and Red have in CESAP. The Ministry of Women has shown interest in providing funds in the near future for some of their projects, but this requires a bureaucratic procedure which takes time. Despite willingness and hard work, their impact on solving environmental or even women's problems have been insignificant: the topics they have chosen are narrow in scope and do not transcend the interests of women's audiences. Similarly with Círculos Femininos, the 'domestic' nature of their environmental demands do not have the capacity to become symbolic political facts through the mass media.

COFEAPRE

The Comisión Femenina Asesora de la Presidencia de la República (COFEAPRE) was created in 1990 by the President of Venezuela. Strictly speaking, it is not an NGO and for this reason it is not a member of CONG. It is composed of parliamentary and high-ranking professional women belonging to different political parties who advise the Presidency on women's issues through the Ministerio para la Participación de la Mujer en el Desarrollo (Ministry for Women's Participation in Development). In the past, they lobbied to get greater representation in parliament and in the decision-making bodies of political parties. COFEAPRE does not define itself as an environmental group, but in 1991 they organized the Primer Encuentro Binacional Colombo-Venezolano sobre Mujer y Ambiente (First Binational Columbian-Venezuelan Meeting on Women and the Environment) with the purpose of formulating a 'new formative model oriented to understand the demands arising from the diverse roles women have to play in today's society'.[24] For the first time in Venezuela, more than 2,000 men and women from different socio-economic backgrounds came together for an event that was considered a great success by all environmentalists – women and men.[25]

While the event was fully organized by women, it was attended by a large number of men from environmental groups, particularly organizations classified as political-ideological such as FORJA. People from the political parties that were members of the Congressional Commission on Environment also attended this meeting.[26] Of particular significance was the attendance at the Encuentro of men belonging to environmental organizations led by men (for example, FORJA) at a time when such groups were weak and needed a larger audience, new members to revitalize the movement, and political legitimacy and power. Noticing the political importance of the heterogeneous constituencies gathered for such a meeting, male environmentalists tried to take control. Nonetheless, they were unable to achieve this due to the fact that their old 'traditional political' strategies and rigid style of leadership did not appeal to the more flexible feminine audience and organizers.[27]

COFEAPRE was created by the President of Venezuela, as a presidential consultancy commission on 'feminine' matters, therefore its objectives are wider than the environmental ones. Noticing the success achieved and the potential that both 'environmental and women's issues' seemed to have to mobilize women and men, the coordinator and main organizer of the Encuentro decided to create the Asociación Venezolana de Mujer y Ambiente (AMAVEN).[28]

AMAVEN

AMAVEN was formally constituted in November 1991 and immediately accepted as a member of CONG. Its objectives were the following: to create environmental consciousness among the population which will allow participation and contributions to the conservation of the environment, with the purpose of improving standards of living and to obtain common and equitable benefits; to promote and defend development, protection and conservation of the environment through the mobilization of all sources of information and dissemination; to cooperate in the solution of environmental problems and to stimulate sustainable development through the active participation of women; to consolidate the democratic processes of social justice, the preservation of peace and the respect of human rights (AMAVEN 1991). As a leading founder put it, 'in this organization, women are supposed to be protagonists in the conservation and defense of the environment'.[29]

Most of the Committee members of AMAVEN also participate in political parties and they know each other from this and from participation in CONG, all of which facilitates the existence of social bonds and friendships among members.[30] Nonetheless, and probably reflecting the structure of political parties, the organization has a more rigid and bureaucratic structure than other women's organizations such as GEMA. This rigidity could have two consequences. It could work against the emergence of an 'invisible network' that activates itself, or surfaces, when there is a problem. This could become a major drawback as the state does not recognize 'invisible networks' and consequent participation as political. Since the stated objectives seem to be general enough to transcend the domestic domain, a formal organizational structure could help to overcome the weak articulation women have with the state and contribute to the recognition of AMAVEN's actions as political. However, the first task is to overcome the potential conflicts that derive from the membership of some members in political parties.

The fact that several of the members belong to political parties could be a potential source of conflict inside CONG,[31] especially if the political parties try either to neutralize or to coopt CONG's actions in view of the increasing importance they give to symbolic themes, such as the 'environment', to revitalize their own precarious political legitimacy. On the other hand, some members of the Executive Committee of AMAVEN are very active in political parties and it is not clear how their alliances would change in the case of conflict between parties, CONG and AMAVEN.

Another potential source of conflict derives from the fact that AMAVEN competes with other women's organizations and with CONG for funds and recognition. By incorporating environmental demands, which are 'fashionable' today, and by creating a formal structure under the umbrella of CONG, AMAVEN has put itself in a more favourable position than other organizations to obtain funds and recognition. Due to the scarcity of power and funds, competition between organizations could make it more difficult to achieve consensus within CONG. Finally, the multiple participation of members of AMAVEN in women's and non-women's organizations could create conflicts of interest and lead to the fragmentation of women's groups, particularly those within CONG.[32]

THEORETICAL CONSIDERATIONS

Women's approach to the environment from a gender domestic perspective is imbued with mobilizing political potential. The Primer Encuentro Binacional Colombo–Venezolano sobre Mujer y Ambiente (First Binational Columbian–Venezuelan Meeting on Women and the Environment) organized by COFEAPRE has shown that when women assume leadership roles they have been able to mobilize both women and men. However, the concretization of this politics into formal or informal organizations, and the transformation of environmental demands associated with women into 'political demands', present substantial contradictions that could reduce this mobilizing potential.

Some of the so-called environmental problems of the poor, such as chronic water shortage and disposal of solid wastes, have been traditionally associated with women's daily routines: the lower the income of women, the stronger this association holds. These tasks are considered part of the 'domestic' domain (MARNR 1991: 17–19). In Venezuela, these daily responsibilities are being officially ascribed to women through institutional environmental programmes. One of the environmental projects of the Ministry of Women specifically addresses the education of female leaders in skills necessary 'to collect and recycle solid wastes in order to improve the quality of life in their settlements as well as to improve their economic situation' (ibid.). This programme also intends to develop alternative technology for the production of food through recycling, small enterprises, cooperatives, environmental conscientization and education. From a feminist point of view, allocation of these tasks could be seen to perpetuate women's traditional role instead of contributing to greater economic equality with men. Further, they tend to promote the perpetuation of women's traditional domestic roles working against the advancement of women's equality. But, it is not just the Venezuelan government: international agencies – the United Nations, the Inter-American Bank and the World Bank – also allocate this role to women in relation to their role in sustainable development. One United Nations document notes that:

> women are critical actors in the strengthening of a rational environmental administration. They are closer to nature since they have assumed traditionally the task of reproduction, not only of the species but of daily life through domestic work. If they are responsible for the socialisation of children and the administration of the resources in their surrounding environment, it is necessary to strengthen this role, via financial and technological help.
>
> (Anonymous 1991)

At the national and international level, the implicit ideology behind this role assignation is 'ecofeminism' (Rao 1991). Women's reproductive role and identification with Mother Nature lead inter-governmental agencies to assign women a greater responsibility for environmental conservation and Nature's protection (ibid.). Women are also responsible for survival, providing water and other domestic tasks for the whole family. Ecofeminism elides two propositions: first, an identification of women's role with their domestic tasks, and second, the assignation of symbolic meaning and political power to a combination of environmental and gender demands. While the second proposition could be applied more properly to Europe and the United States, in Venezuela, as a feminist recently stated, 'very few women understand Ecofeminism, much less its political implications'.[33] This lack of information and a political perspective reduce the potential of ecofeminism to create a new political fact from the environmental.

From the experience of Círculos Femininos and Red, we could conclude that when environmental problems are identified with the domestic domain those problems have little, if any, 'symbolic' value. Consequently, it will be more difficult for such problems to become political and to gain attention from the government. However, it is not only the identification of the 'environment' with the domestic or the political domain that influences the success or failure of transforming the 'environment' into a political fact. Organizational structure, strategy, relations with political parties and the use of the mass media also become very important in explaining the success of an environmental organization. Strategies that favour the conversion of the 'environment' into a political fact seem to be associated with gender: men are more able than women to convert even domestic problems into what are traditionally defined as 'political problems'. In contrast, women leaders of an organization are more able than men to transform environmental-non-domestic problems into 'new' political facts.

The experience of AMIGRANSA shows that the capacity to convert women's environmental problems into 'new' political facts depends first upon the definition of the demand as 'non-domestic' and second upon its transformation into a political fact which has a symbolic power and is independent of political parties. This is done largely through the mass media. Following this argument, we could say that CONG has been unable to permeate Venezuelan political culture with women's demands for several reasons. Women journalists, who collaborate with CONG, are instructed to write about women's topics – sexual discrimination, female work, children, poverty and the family – that have little, if any, symbolic value *per se* for the media. On the other hand, the fact that those issues are printed in special women's sections of the newspapers greatly reduces women's access to the political potential of the printed media. In women's environmental organizations, the subordination of the environmental to gender demands defined within the domestic domain also reduces this potential (Rao 1990).

WOMEN, ENVIRONMENT AND NEIGHBOURHOOD ASSOCIATIONS

The urban movement or neighbourhood associations has been very strong in Venezuela since the 1970s and they have proliferated throughout the 1980s (Santana and Perrone 1991). Identified primarily with the urban middle class, neighbourhood associations also emerged in squatter settlements under the tutelage of the political parties. According to Ley Orgánica del Regimen Municipal (Organic Law of Municipal Government), neighbourhood associations are the repositories of ecological concerns in the broad sense of the 'quality of life'. As in Mexico, a characteristic feature of the Venezuelan urban movement is that it developed out of the neighbours' movement as part of the movements *de cuadro de vida* (for quality of life) (Ovalles 1987). One of the first such movements, the Movimiento de Integración de la Comunidad (MIC: Movement for Community Integration), had among its objectives alternative proposals of a clearly ecological nature, including alternative energy resources, recycling and the control of pollution at the community level through community organization.

Since the late 1970s, neighbourhood associations have become the articulating axis of social mobilizations for socio-political and economic democratization and decentralization. The neighbourhood movement succeeded in integrating cooperative, popular, environmental and even feminist organizations around the issue of *democratization* – the issue with the most symbolic effectiveness and mobilizing potential during the 1980s (Uribe and Lander 1991). As a result of this integration, the 1980s witnessed numerous joint actions by neighbourhoods and environmentalists aimed at addressing not only participatory and social democracy, but also environmental and/or urban issues at the local, regional and even national levels. This collective orientation towards 'participatory' democracy was very strong in the 1980s when it had a 'modernizing' effect on political culture. Given its symbolic value, both men and women participated actively in this process but the leaders of the movement were primarily men.

In general, we can argue that women are more deeply rooted in their communities than men due to their role as housewives and mothers (Logan 1990). Women's motivation to mobilize often rests in their self-definition as mothers and their commitment to fulfilling the responsibilities implicit in this role. In Venezuela, women participate more than men in organizations that demand public or collective urban services such as water, schools, parks, pavements (Ray 1969).[34] The emergence of urban neighbourhood associations in Venezuela in the 1960s was due to a group of middle-class women who organized themselves against the location in their neighbourhoods of activities they considered would lead to a deterioration in their quality of life (Lope Bello 1979). With the promulgation of the Ley Orgánica del Regimen Municipal of 1976 neighbours' associations flourished in middle- and low-income neighbourhoods where women and men participated. But, in spite of the high level of activism among women, most of the formal leaders of the Federación de Asociaciones de Comunidades Urbanas (FACUR: Federation of Urban Community Associations), the Escuela de Vecinos (Neighbourhood Schools) and the Coordinadora Nacional de Federaciones de Vecinos (CONVECINOS: National Coordinator of Neighbourhood Federations) have been men.

The economic crisis of the 1980s has contributed to a differentiation of urban demands according to social class and gender, affecting women from the poorest settlements the most. The worsening of material conditions of the poor has brought together urban demands with the environment. The most important demand of the impoverished population is basic services while for the middle-class it is the defence of the living conditions already achieved, at the expense of the lower class, since both are competing for underfunded basic and social services. Thus, the economic crisis has accentuated class antagonisms in the cities and rendered 'visible' the acute division by social class.[35]

Another consequence of the crisis is the differential impact it has had on men and women. Women tend to perceive the material conditions of their neighbourhood as obstacles to the proper performance of their motherhood role (Logan 1990). Therefore, the economic crisis has affected more women from the low-income settlements than men, because they are responsible for the household economy and family survival. As the economic crisis and the cuts in social welfare threaten the very survival of increasing proportions of squatter settlements' populations, many neighbourhood associations sought solutions to this impoverishment. In this process, women played a more important role than men. Consistent with women's traditionally defined roles, women in the squatter neighbourhoods took on the responsibility of providing the basic necessities of life without rebelling against prescribed gender roles. As Saporta, Navarro-Aranguren, Chuchryk and Alvarez have noted, 'In keeping with their socially ascribed responsibilities as "wives, mothers and nurturers" of family and community, women have taken the lead in the day-to-day resistance strategies of Latin America's popular classes' (Saporta *et al.* n.d.) Therefore, the most compelling problems around which Venezuelan women mobilize fall into what we could define as the domestic domain of family and community. For these mobilizations, they optimize the links established among themselves through overlapping networks of kinship, friendship, neighbourliness and compadrazgo (Lomnitz 1975).

Although women are more visible in the squatter settlements in their fight for survival, they do not have any political power. The political representation of the community when dealing with government or private institutions is assigned firstly to men. Given this lack of political power and their restricted access to the mass media, low-income women do not have the capacity to permeate the political system by transforming their demands for survival or basic services into a powerful political issue. Moreover, as a result of some legal measures taken to counter the political crisis, Venezuelan women from all social classes lost the possibilities of access to political power in municipal or local participation when, in 1989, a new Reglamento para la Participación de los Vecinos (Neighbourhood Participation Regulation) was added to the Nueva Ley Orgánica del Regimen Municipal (NLORM). This transferred political power to neighbourhood associations in local or municipal matters (García Guadilla 1991a). These legal developments had several relevant consequences.

As opportunities to mobilize decrease due to the economic and political crisis, more men actively look for leadership positions inside the neighbourhood associations. This provides men with a route whereby they can enter local and municipal politics. The political parties are now eager to coopt and intervene in neighbourhood associations in contrast with their past lack of interest. This has accentuated the conflicts inside the neighbourhood movement, between the Federación de Comunidades Urbanas (FACUR: Federation of Urban Community Associations) and among the Confederación de Federaciones de Vecinos

84

(CONFEVECINOS), for example. The combination of the above trends could lead to a redefinition of the existing 'invisible networks' through which women tend to participate (García Guadilla 1991c, 1992), and to their cooption and absorption by the state. Ultimately, it could discourage women from participating in local municipal politics.

CONCLUSION

Women's political strategies in relation to environmental goals differ from men's. Women's environmental groups mediate their demands through gender and tend to identify them with the role of motherhood and survival, while men organize their demands around economic and political issues. Even when men mobilize around survival problems and basic services, they tend to identify those issues as political and non-domestic.

As I suggested at the beginning of this chapter, by way of their ascribed roles women are linked to political power in weaker ways than men. Governments do not recognize the domestic domain and demands derived from this domain as 'political' and they are unable to understand the political significance of the 'invisible network' women have built to achieve their aims. They are 'invisible', of course, not for women but in relation to the extant political structures.

Given the difficulty women have had in entering traditional politics through the social movements, some women created 'alternative' or 'new' spaces of political significance around environmental or ecological problems that could have a greater potential to mobilize people than traditional politics. Women's ability to create this new political space, and consequently new political facts, seems to be greater than men's and it is correlated with the factors listed below.

Environmental demands have more symbolic value than women's issues or urban agendas, but when they are identified with the domestic domain of women they tend to lose this potential. Organizational structure, strategy and leadership style play an important role in the generation of symbolic effectiveness and in the success of social movements. The power of an organization rests not on its size but on its ability to gain wider articulations with other organizations. Those organizations with a more flexible and horizontal structure have greater potential to mobilize other organizations around their own demands. The afore-mentioned structural and strategic characteristics vary greatly according to the gender composition of the leadership.

First of all, organizations where women participate as leaders tend to adopt 'negotiational' strategies, be more pro-active than defensive, and effectively use institutional channels, personal relationships and the mass media. Second, strong social bonds facilitate consensus within groups and reinforce organizational flexibility, stability and adaptative strategies which are essential for success. These characteristics are more common among the women's organizations born out of primary social relationships between members. Secondary social relationships that characterize, according to the theory of social action, 'voluntary or interest' organizations such as the ecological, do not, in women's groups, exclude the primary relationships. In women's environmental groups they tend to coincide and constitute an important factor in the success of their strategies. Finally, 'invisible networks' are characteristic of women's participation. Political parties are eager to appropriate new political meanings once the old slogans do not mobilize people. However, the amorphousness and occasionality of women's organizations reduce their visibility, making cooption by the government and the political parties more difficult. Women seem to value autonomy from the political parties more than men and they are more willing to give spare time and resources without getting any political reward in their struggles to build political facts that differ from the traditional ones generated by parties.

Women in recent Venezuelan history have been excluded from traditional politics and power relations associated with the political parties and tend to identify those parties with '*machista* politics'. Therefore women are more willing to express their demands through alternative political spaces where some new organizations are being created although it is not yet perceived by or 'visible' to the government or political parties. The media play a decisive role in the generation of the new political culture. During the 1970s and 1980s women more than men tended to use the mass media to transform Venezuelan political culture and recreate the 'environment' as a political fact. It is through the media that the 'environment' is consumed, produced and transformed into a new symbolic and political fact. Finally, social class, professional and educational background and other individual characteristics tend to accentuate some of the above tendencies.

It is clear that, to be effective, all the factors have to come together. If one of them is missing, demands will not be transformed into a political fact. AMIGRANSA has been more successful than the other environmental organizations in achieving its objectives. Its success stems from the fact that its themes and values generate new, nationally or internationally relevant political facts, independent from the traditional parties. Nonetheless, it is not only the strategy that has generated this success, but also the symbolic value the ecological message has. While the political-ideological and the symbolic-ideological organizations share the same demands only the latter organizations create political effectiveness through their strategies, structure and style. On the other hand, women's environmental organizations tend to pursue strategies similar to AMIGRANSA, but their demands lack symbolic value due to their identification with the domestic domain.

The possibility of transforming women's environmental domestic problems into symbolic ones, relies on rethinking such issues as global problems that affect men and women alike. This global conception of the 'environment' should transform problems traditionally associated with the domestic domain into political problems, thereby attracting government's attention and stimulating wider alliances and articulations among different social movements. The Global Assembly of Latin and Caribbean Women on Women and the Environment, held in Miami in November 1991, pointed in the right direction by taking a global approach toward the 'environment', and addressing the economic order as the cause of the environmental problems that affect women and men equally and by proposing general solutions to achieve sustainable development. The conference called for male and female journalists to join in the efforts to protect the environment and to use the mass media more effectively in relation to a new 'eco-politics'. It is a politics in which women will be foregrounded.

4

'WE LEARNED TO THINK POLITICALLY'

The influence of the Catholic Church and the feminist movement on the emergence of the health movement of the Jardim Nordeste area in São Paulo, Brazil

Leda Maria Vieira Machado

INTRODUCTION

There was a change in the system of government in Brazil in the mid-1960s which brought about a period of political repression, in which all forms of popular reaction and organization were either eliminated or severely restricted. From the mid-1970s onwards this situation started to change. Among other things, a series of urban movements started to spring up. Contributing to this process of change were the Popular Church (a sector of the Catholic Church in Brazil), and in the specific case of the participation of women in urban movements, the Feminist Movement. It should be noted that there is a distinction between the Feminist Movement itself, comprising the several feminist groups in Brazil, and the overarching ideology which in various forms, and with various emphases, was promoted by these groups. Here only the ideology of this movement is considered, for in the case study under scrutiny no feminist group was present.

This chapter examines one movement, the Health Movement of the Jardim Nordeste area, to illustrate these influences. It takes place in São Paulo in the south-east of Brazil, the most economically developed region of the country. The great diversity of Brazil, both in geographic and ethnic terms, should, however, be borne in mind. It has a sharp contrast between rural and urban areas, as well as between regions, as the examples of Amazonia in the north and the metropolitan area of São Paulo in the south-east testify.

This chapter tackles two questions. First, how did the Catholic Church, particularly the Popular Church, contribute to the formation of one women's group, the Health Movement of the Jardim Nordeste area? This question is fundamental for understanding the development of this movement and women's contribution. Due to the level of political repression in the 1970s in Brazil, one of the only options open to women for participation in life outside the home was in the Catholic Church, usually in mothers' clubs. Within these parish-based clubs, the women of the Jardim Nordeste area[1] began to develop a more political consciousness, which led them to try to find more effective ways of solving the concrete problems that affected their neighbourhood, such as the lack of any form of healthcare. The second question addressed here is to what extent did the Feminist Movement in Brazil, with its specific characteristics, contribute to women's mobilization for better healthcare in the area? This contribution was more subtle, depending on the spread of feminist ideology among low-income women, which played an important role in legitimizing their participation in urban movements in Brazil, especially from the mid-1970s onwards. The influence of feminism on women's movements has been so clearly detectable, not only in Brazil but also throughout Latin America, that one of the main issues today in the literature on gender and urban movements is to what extent movements in which women take part are *feminist* (they challenge the division of labour along lines of gender) or *feminine* movements (they reinforce the division of labour along gender lines) or whether the latter have the potential to become feminist. Apart from the fact that these definitions are still very much open to debate, this approach underestimates the *process* of mobilization, for it concentrates on the end results, and as a consequence ignores the changes in women's lives which have fundamental implications for how gender is perceived and constructed. This highlights the need for a historical approach, so that this can be avoided, but still stressing the significance of gender on urban studies.

This chapter is divided into four parts. The first is a presentation of the overall socio-economic and political conditions in Brazil in the 1970s and early 1980s, as a background to the discussion that follows. The second covers the Popular Church in Brazil and in São Paulo, with a view to describing the evolution of this progressive sector of the Brazilian Catholic Church. The third is about the Feminist Movement in Brazil and in São Paulo, describing how it has developed and how it came into contact with urban movements. And the fourth is on the emergence of the Health Movement of Jardim Nordeste and how the Popular Church as well as the ideology of the Feminist Movement influenced it. The assertion of this chapter is that both elements were decisive for the emergence of the Health Movement.

BRAZIL IN THE 1970s AND EARLY 1980s, AND URBAN MOVEMENTS

A right-wing military *coup* took place in Brazil in March 1964. This radical move was a reaction by the armed forces to the institutional

89

deadlocks and dilemmas produced within the government apparatus itself. There had been growing popular dissatisfaction, expressed in urban mobilizations and struggles, and the government had been unable to deal with the social and political turmoil, throughout the terms of Jânio Quadros (who resigned in August 1961 after just seven months in office) and João Goulart (October 1961 to March 1964) (Cardoso 1979).[2] After the *coup*, a two-party system was installed, with a government party (National Renewal Alliance – ARENA) and an opposition party (Brazilian Democratic Movement – MDB), and abolition of the previous multiparty system. As one military government succeeded another it gradually became obvious that one of the devastating effects of this change in the political structure of the country was an increasingly sharp reduction in living standards for the majority of the Brazilian people, as the direct result of a policy of income concentration. This was based on very low wages for the working class and higher levels of capital accumulation, as well as an emphasis on the production of consumer goods. It gave the new regime support, particularly from the middle classes. A process of rapid industrialization and modernization of rural areas was started by the new regime. Within a context of income concentration, the Brazilian economy grew at one of the fastest rates in the world between 1967 and 1980 (Viola and Mainwaring 1985: 201). This economic success yielded support for the government from both upper and middle classes which benefited directly from this expansion.

At the same time, a process of impoverishment of large sectors of the population occurred, reflected in the fall in the levels of real minimum wages.[3] Called '*arrôcho salarial*' (wage tightening), it began under the first military administration in 1964, and this situation gradually worsened with a consistent and growing process of migration from the countryside to major towns, where the majority of migrants found no place either in industry or in the service sector. In 1970 the wealthiest 20 per cent of the Brazilian population owned 62 per cent of national income. In 1976 this group had increased its ownership to 67 per cent of national income. By contrast, in 1970, the poorest 50 per cent of the total population owned just 15 per cent of the total income of the country, a figure which had been significantly reduced to 12 per cent by 1976 (Kowarick 1982: 31). Figures from the São Paulo Metropolitan Area[4] are a striking example of how unevenly people were paid under the military regime. Despite the volume of income production in the Greater São Paulo area (the largest industrial concentration in Brazil, responsible for more than half the income of São Paulo State, and nearly a quarter of that of the whole country),[5] a large proportion of its population received very low incomes. In the municipality of São Paulo alone the minimum wage fell by about 55 per cent between 1959 and 1979 (ibid.: 32).

These circumstances particularly afflicted Brazil's low-income groups, and at the beginning of the 1970s prompted demonstrations and urban movements around issues of consumption such as water, transport, cost of living, crêches and sewerage. A dominant author in this area of the literature on urban studies has been Castells (1977, 1979, 1983). His work on the 'collective means of consumption' (services and goods which by definition are provided by the state as a fundamental element in the process of reproduction of urban labour-power) shifted the focus from production to consumption. For him, conflicts, meaning urban movements, arise when these means of collective consumption are not provided. In his latest work, *The City and the Grassroots*, he mentions the *decisive role of women* in his investigations (Castells 1983: 68). However, this is misleading. Gender is neglected in this work. Castells' only, though quite explicit, mention of gender was when he stated: 'It is our hypothesis that there is some connection between the social character of urban studies and the role of women in these movements' (ibid.). Notwithstanding his important contribution in this line, Castells was immediately concerned with the problems and difficulties associated with the transition of women's participation in urban movements and women's liberation. His studies were, from the outset, limited by a single alternative: they were about whether and why women's movements, movements with a women-based character, could or could not transform themselves into feminist movements. Because the end result of his investigations in this respect was that gender relationships were not at the core of any movement, despite the decisive participation of women, nothing much else could be done in terms of research in this topic – as though the redefinition of the urban meaning in this respect could be achieved only through feminist demands. The discussions under the headings below also aim at contributing to this area of research, stressing the importance of gender for the study of urban movements.

Notwithstanding the complexities of Castells' propositions, they offer an important perspective on developments in Brazil in the late 1970s and in the 1980s, particularly in the light of the observations of Evers *et al.* (1982) and Moser (1987), that women, especially in urban areas, are always more involved in neighbourhood organization over consumer issues, especially in relation to the provision of collective means of consumption, for they are the ones who suffer most and take responsibility for the survival of their families and communities. Gender roles and deteriorating living conditions alone, however, are insufficient to explain the spontaneous upsurge of urban movements within the country (Kowarick 1985). Besides the increasing impoverishment of the poorest sectors of the Brazilian population, the totalitarian nature of the Brazilian state also contributed to the readiness of the poor to seek new ways to express their discontent and demands. According to Cardoso

(1983a), the 'suppression' of institutional channels of expression contributed to the direct and 'authentic' action of the popular sectors of the population.

General Geisel took office in March 1974 (until February 1978, the fourth military government since 1964). This takeover was due basically to a sharp increase in international oil prices, which meant that the economy no longer had enough resources to include all the middle-income sectors of the population, which were one of its major sustaining pillars. Moreover, increasing economic pressure was being put on the poorer sectors of the population, yet the political regime installed in 1964 faced internal difficulties and was unable to exert hegemony over society as a whole (Cardoso 1979). Within this context, the Brazilian population started to express its anger towards the military regime and its violations of individual liberties, thus establishing a clear gap between the government and the people, and revealing the regime's lack of legitimacy. This also signalled the beginning of a struggle for the re-establishment of a democratic regime (Telles and Bava 1981; Velasco e Cruz and Martins 1983; Jacobi 1983). As the revolt intensified, against a background of stringent repression, the conditions were created to unite many institutions, associations and movements in acts of protest, setting the scene for the establishment of an 'informal alliance' between democratic opposition parties (Velasco e Cruz and Martins 1983). The situation became unsustainable for the regime when certain Brazilian businessmen themselves contested the 'legitimacy' of the government and joined in expressions of discord by actively opposing moves by the new regime to nationalize the Brazilian economy (Cardoso 1986). These expressions of discord already counted on the unions, which for the first time since the military *coup* expressed their political opposition to government attitudes and directed a campaign for wage restoration (Velasco e Cruz and Martins 1983). The Brazilian population engaged in a process of reorganization and articulation as the military slowly relaxed its control. In a new context without space for government tutelage, the population began to establish ways of communication with government offices, and of challenging them (Cardoso 1983a).

However, all these processes of mobilization which started in the mid-1970s were impeded by the fact that in the immediate past, Brazilian society had seen the complete destruction of any form of organized popular expression. All that was left to build on were small groups such as mothers' clubs, local associations and grassroots communities of the Popular Church. In terms of formal politics, the first results of this work came to light in the November 1974 Congressional elections, which were won by the only opposition party at the time (MDB), in spite of government efforts to manipulate the polls. Win or lose, the military government still dominated the political sphere and had the initiative.

Since it had been successful in suppressing the radical opposition, the regime could afford to introduce policies which had a more 'open' content and conveyed an image of liberalism on which the military capitalized (Viola and Mainwaring 1985). Among these were the gradual reduction of the control over freedom of speech and a partial amnesty for political prisoners and exiles. However, the most significant move was the reinstatement of a multiparty system, making it possible for opposition parties such as the Partido dos Trabalhadores' (PT: Workers Party), the Partido do Movimento Democrático Brasileiro (PMDB: Brazilian Democratic Movement Party) and the government party, the Partido Democrático Social (PDS: Social Democratic Party), to appear on the political scene. General Figueiredo became the fifth military president in March 1978 and gave continuity to the process of change initiated by Geisel, which later became known as *Abertura* (political opening up of the regime). Nevertheless, the military continued to produce authoritarian laws like the *Estado de Emergência* (State of Emergency) which provided the president with powers to cancel all civil liberties, to intervene directly in unions and to judge in military courts.[6]

Throughout this process of change, the standard of living of the low-income population consistently worsened. The number of those earning more than a minimum wage was reduced in 1979 (Lagoa 1985: 57) whereas the number of those unemployed increased. In São Paulo the industrial sector employed 2.1 million people in 1980, a figure which was reduced to 1.6 million by 1983, and in the course of these two years some 500,000 workers lost their jobs (Betto 1984: 11–12). The price of consumer goods rose by 155 per cent from 1982 to 1983 in contrast to a 112 per cent increase in wages (ibid.). A worker who received the minimum wage in 1950 had to work thirty-two minutes to purchase a kilogram of beans, but three hours and twenty minutes in 1983, so he or she was six times worse off (ibid.).

As a result, urban movements continued to spring up and some, such as the Health Movement of the Jardim Nordeste area, were even strengthened. In most urban movements women made up the majority of participants. It is estimated that of every 100 participants 80 or more are women (Rede Mulher 1985). From the end of the 1970s, the country witnessed an upsurge of workers' and urban movements, local groups, professional associations of all kinds and the *Comunidades de Base* (grassroots communities) supported by the Popular Church. Organized urban movements – especially in the larger centres – had started around issues of consumption in the poor neighbourhoods and gone forward, working for an improvement in living standards (Brasileiro 1982), which transcended the neighbourhood sphere. Issues of consumption included infrastructure and services such as nurseries (Gohn 1985; Jacobi 1983), healthcare (Jacobi and Nunes 1981; Machado 1988;

Corcoran-Nantes 1988), improved transport (Telles and Bava 1981), slum upgrading (dos Santos 1981; Castro 1983; Diniz 1983) and land tenure (Jacobi 1983). Issues of living standards included campaigns against the rise in the cost of living (Evers 1982), the Slum-dwellers' Movement (Corcoran-Nantes 1988; Boran 1989) and the Unemployed Movement (Corcoran-Nantes 1988). Several more spontaneous mobilizations also took place such as riots for better transport (Nunes 1982), looting of supermarkets (Barreira and Stroh 1983) and riots of construction workers on the building site of the underground in Rio (Valladares 1981).

This whole process of change in Brazil during the late 1970s and early 1980s, in which women were ever present, was thus the result of a conjuncture of political and economic factors. There were, however, two other factors which contributed to this process: the Popular Church and the Feminist Movement. These will be examined below.

THE POPULAR CHURCH IN BRAZIL AND IN SÃO PAULO

The Popular Church as it is known today in Brazil was generated after the military *coup* of 1964. During the Second Vatican Council in 1965, Brazilian bishops gathered in Rome to approve measures to encourage a rapprochement between the Catholic Church and large sectors of the people, through the attraction of lay people to work with the Church close to the community. This contact in Brazil had been damaged not only by the support given by the Church to the military *coup* for fear of the advancement of Communism in the country, but also, more importantly, by the appearance of new religious forms like Kardecism (a branch of the Catholic Church concerned specifically with spiritualism) and the Pentecostal Church, which were growing and achieving greater support among the population, particularly the poor. The new popularization of the Catholic Church also reflected the Latin American Bishops Conference (CELAM) at Medellin, Colombia, in 1968, which expressed its clear support for the struggle for social justice for the Latin American poor. This transformed the face of the Church in the continent, including Brazil, and was particularly important in clearly defining the new line of action the Church should follow, strengthening the local base communities. Documents from Medellin specified that the Church should have as its objectives the defence according to the Gospel of the rights of the poor and 'oppressed', the encouragement of people's development of their own base communities, and the support to their search for *true justice*. The Church considered the Christian base community to be the fundamental ecclesiastic core responsible for the expansion of faith as well as the primary factor of human promotion (quoted in de Camargo *et al.* 1981).

The first Ecclesiastic Base Communities[7] (CEBs), which originated after 1964 and retain most of the characteristics to this day, gained a new impetus from 1968. Following the 1964 military *coup* a number of bishops had already expressed political opinions which differed sharply from the official position of the Brazilian Catholic Church, whose leadership had been hesitant to condemn the new regime and its oppression. This sometimes transformed entire dioceses (Mainwaring 1986). However, such transformation was not to take place in the archdiocese of São Paulo until 1970 when the conservative archbishop was replaced by a progressive, D. Paulo Arns. Many progressive nuns and priests had been working closely with the working class and they were now able to establish a stronger channel of political activity with the approval of São Paulo's new Catholic Church leader. They had been engaged in the organization of evangelization courses consistent with the objectives laid down by the Second Vatican Council, which aimed at developing a critical awareness among the participants, followed by the development of the ability to work together in solving communal problems. This work laid the foundations for the work of the Ecclesiastic Base Communities within low-income communities (de Camargo *et al.* 1981), to which the new archbishop contributed much. According to Mainwaring (1986), under D. Paulo's leadership, the base communities, the defence of human rights and the rights of the poor became archdiocesan priorities.

This fact gains weight when placed against the oppression and silence imposed on Brazilian society after 1964 and especially in São Paulo during the late 1960s and early 1970s, for it was here that left-wing groups, some engaged in armed struggle, were most active and organized and here that repression was at its strongest. Without a legal channel to convey their demands and aspirations, most Brazilians, especially those from low-income backgrounds, found in sectors of the Church, particularly through the Ecclesiastic Base Communities, the most structured opposition there was to the regime throughout the 1970s (Lopes 1980). The Church was the only institution which was able to attack the economic model, challenge the strategy of repression, defend human rights and organize popular sectors. In this sense, the 'political vacuum' was an incentive for the transformation of the Church.

The Ecclesiastic Base Communities met in people's homes and discussed the content of the leaflets the Church had provided, which told a story about the problems of the poor. They did not have a strictly religious content. Although discussion was followed by talks about the Gospel there was already a political motivation for these discussions, which included issues on the social organization in Brazil. Thus the increasingly political initiatives of progressive nuns and priests, and the participation of lay people in grass roots organizations from the mid 1960s, provided a groundwork for the growing process of popular mobilization, throughout the 1970s, especially for urban and labour movements (see Vink 1985). The Popular Church, as a sector of the Brazilian Catholic Church, particularly in São Paulo, was characterized by its involvement in urban movements and resistance to the military government (see Duarte and Yasbeck 1982). Although crucial at a particular historical moment in Brazil, it must also be noted that the influence of the Popular Church over the development of urban movements was not spatially homogeneous or consistent. In fact in many cases the development of these movements may also be due, among other things, to their having distanced themselves from the work of the Ecclesiastic Base Communities, because their work had gone beyond the first phase of articulating the wants of the people, to which most of those communities were dedicated exclusively (Lesbaupin 1980). This appears to be the case in the Jardim Nordeste area, in which the women who had been taking part in the parish and the Ecclesiastic Base Communities, decided to leave the parish's activities to dedicate themselves to the organization of an independent movement (see p. 105–6).

THE FEMINIST MOVEMENT IN BRAZIL AND IN SÃO PAULO[8]

The other factor that contributed to the process of change in Brazilian society during the 1970s, particularly in relation to the political participation of low-income women, was the Feminist Movement. The Feminist Movement in Brazil, while aiming to influence the mobilization of low-income women on issues such as wages, working conditions and maternity leave, was composed mainly of white middle-class women. Moreover, these women usually met at times when it was difficult for low-income women to attend, and in fact few ever did (Schmink 1981; Cardoso 1983b). Furthermore, the Feminist Movement in Brazil historically tended to direct itself to broader political questions, relatively remote from specific women's issues, which were actually relegated to a secondary position (Schmink 1981) – seen as necessarily linked to a feminist struggle, but only in so far as this helped to mobilize women in their demands for reform rather than revolution (ibid.). Alvarez (1990) describes Brazilian feminism in the 1970s as rather undefined, and largely restricted to 'established' Marxist categories and analysis. The main concern of those involved in organizing women in relation to discrimination was the link between women's struggle and class struggle, and in general women's 'specific' needs were seen as secondary to the need for 'general social transformation', so that the particular would wait until 'after the revolution'. This was also the trend elsewhere in Latin America (see Jaquette 1989). The International Year of Women, promoted by the United Nations in 1975, gave a major impetus to women's groups in Brazil, some of which, like the Centre for the Development of Brazilian Women, counted on the participation of unions, political parties, sectors of the Church and women's groups (Schmink 1981).

There is some disagreement as to the more recent work of feminist groups. According to Schmink (1981), a few feminist groups emerged with the notion that community issues and women's issues were different and specific, but most groups continued to propose a unified view of citizens' issues as a whole. After the Second Women's Congress in São Paulo in 1980, there appeared a distinct feminist political identity within the overall movement, which confronted the traditional link made by the Brazilian Left between women's groups and the overall class struggles, in practice subsuming the former's demands. So the 'generic' earlier activities gave way to more 'genderic' ones (Alvarez 1990). Barroso (1982), on the other hand, states that one of the main characteristics of the new movement in feminist organization was an increasing focus on issues affecting women directly and specifically. Since the emergence of urban movements in Brazil, the Feminist Movement appears to have been instrumental in the general process of popular reaction, largely because of its ability to legitimize the participation of women in urban movements (Caldeira 1987). It did this by stimulating women to deal with women's problems by redefining public and private spaces and gender roles. Women's low-income neighbourhood groups received direct help from feminist groups, particularly in discussing questions such as family planning, working rights, everyday sexual violence and maternal health (Barroso 1982). This was especially the case in São Paulo, where feminists organized three state-wide Congresses of Women, with an emphasis on women's issues. These congresses included women from Church groups, mothers' clubs from the periphery of the city of São Paulo and other municipalities, and women's groups (whose members were also union members and included feminist demands in their journals and meetings). Discussion ranged over subjects like the maintenance of crèches, neighbourhood demands, the emancipation of women from domestic work and the transformation of such work into specialized activities in the division of labour.

Two groups in particular had pre-eminence in São Paulo. The first, 'Nós Mulheres' (We Women), published a journal of the same name, and the other, called 'Brasil Mulher' (Brazil Woman), also published a periodical under the same name (Barroso 1982; Alvarez 1985, Singer 1981). Barroso (1982) suggests that the representation of feminist groups in São Paulo was stronger than in any other part of the country possibly due to the greater numbers of women incorporated into the salaried workforce and the work developed by feminist groups themselves, particularly among low-income women. Despite problems in terms of different lines of work to be followed which developed between feminist groups from 1981 (see Sarti 1989), the overall ideas and propositions of the Feminist Movement were widespread, even though this movement was not directly active in every district in São Paulo (Caldeira 1987). In Jardim Mirian in the Southern Zone of São Paulo, for instance, a feminist group was directly involved in the area. As far as their direct intervention is concerned, the action of these groups was very localized. In the case of Jardim Nordeste, what was present physically were the CEB's and the Catholic Church, which were not active in the same direct way in other areas of the city. So what takes place has uneven effects throughout the city and each case should be examined on its own (see Alvarez 1990).

THE EMERGENCE OF THE HEALTH MOVEMENT OF THE JARDIM NORDESTE AREA

The district of Jardim Nordeste was founded in 1821, but it was only in the 1940s that it really started to take shape and develop, and to be 'engulfed' by what is referred to in this chapter as 'the Jardim Nordeste area'. The area, located by one of the major railways and close to one of the main highways from São Paulo, which go to Rio de Janeiro, started as a stopping point on this route. The Jardim Nordeste area has historically been a residential district with few commercial activities. In 1940 the area had a total population of about 18,500 (Langenbuch 1971: 171) which grew to about 152,000 by 1970 and to 242,000 by 1980 (EMPLASA 1982: 57). It is a low-income area, where in 1980 only 40 per cent of the total population aged 10 years or over were economically active. Even considering the significant number of children between 10 and 17 years of age attending school, about 11 per cent (ibid. 530–1), and allowing for optimistic estimates for the so-called informal activities (which of course existed), this figure reveals a high rate of unemployment or at best very low and irregular incomes. In 1977, about 70 per cent of the area's economically active population were on low incomes of up to US$400 or £190 per month (Rolnik et al. 1989: 65).[9] It is significant to point out that in the Metropolitan Region of São Paulo between 1978 and 1986 women's incomes were on average less than half of those earned by men (Brant et al. 1989: 51), and the literature gives no reason to believe that this would be very different in the area of Jardim Nordeste. This is also an area in which there are practically no illegal settlements today, but in which piped water supply is partial (as is power), and in which there is virtually no sewerage (Rolnik *et al.* 1989: 67).

The Health Movement of the Jardim Nordeste area was organized in 1976 with an aim to improve the healthcare conditions in the area.[10] Their first demand was for a health centre, then equipment and doctors for the centre. Later, it was for a larger health centre, which was built. From then on, the functioning and equipment of the health centre were always on the agenda of demands. It started with five women and at some points it came to have about twenty women. Throughout the development, though, it has always been some four women who were constantly active and who have carried it on, who counted nevertheless on the support of the women of the neighbourhood when numbers made a difference such as in convoys and demonstrations. Since it has been part of the overall Health Movement of the Eastern Zone of São Paulo, demands have been broadened to include even a say in the allocation of resources as well as in the design of health policies and programmes and in the structure of healthcare provision.

All five women who started the movement came from the mothers' club of the local parish; this has always been an entirely female movement. One of them has since died, and another moved away, while the three remaining are in the Health Movement to this day. It is relevant to give a brief description of these women's life histories so that an idea of who they were can be depicted. (The names given here are pseudonyms.) All three women were born outside the city of São Paulo. Two came from the north-east of Brazil (Zelia and Monica), and the third (Zilda) from the interior of the state of São Paulo. When they first moved to São Paulo, they initially went to another district of the Eastern Zone, and only later moved to Jardim Nordeste where they bought houses along with their husbands. By the time they started to mobilize and to organize the Health Movement, however, they had already been in the neighbourhood for a considerable time: Zelia for twenty years, and Zilda and Monica for sixteen years. Two of them had worked before they married, Zilda in a textile factory full-time and Monica in a hospital kitchen. After they married, both of them initially worked at home, doing sub-contracting and piece-rate work (cutting rubber sandals, cutting underwear, making small plastic or cloth bags, sewing for small-scale workshops) (see Abreu 1986). Zelia did not start work until her husband died, and she then worked as a daily cleaner in family homes until her children were able to work and support the whole family. Although they were better off than the poorest in the area they were still low-income. All three women had children – Zilda and Monica each had four, and Zelia had seven. While the women were literate, none of them had gone further than primary school. Zilda got as far as the fourth grade, and both Monica and Zilda as far as the second.

Zilda stopped doing sub-contracting and piece-rate work when her third daughter was born. She had to do so, because her daughter was handicapped and needed special attention. After her daughter died (at 2 years old) she intensified her participation in the local parish, and from then on she never worked again because her husband's wage was felt to be enough to support the family. When Monica's youngest daughter was 6 years old in 1971, Monica started working as a cleaner in the offices of the sanitation company in São Paulo's Public Works Department, but after about two and a half years she had to give up for health reasons. In consequence she was eventually retired, in 1977. Nonetheless, she continued to work, taking care of her neighbours' children, as well as washing clothes for them. The participation of all the women in the Catholic Church was linked to the need they felt to do something outside their homes: 'to fill an emptiness, to be able to get out of the house' (Zilda, 52 years).

Later they stopped meeting in the parish and used their own houses and attended meetings of the Ecclesiastic Base Communities of the parish. When the idea of a struggle for health started, none of them had young dependent children, and none of them was working on a regular basis. Zilda and Zelia were not working at all, whereas Monica did occasional knitting at home for various clients in the neighbourhood. This gave them all the opportunity to be available and to be able to devote time to the development of the Health Movement, which remains a great commitment, involving the organization of demonstrations, bazaars, jumble sales and meetings. It is also significant that these women have never had insurmountable problems with their husbands which might have hampered their participation in the Health Movement – although this is not universal among husbands in the area. Zelia was already a widow when she became involved in the Health Movement, but she believes that she would probably have had difficulties with her husband concerning her participation had he still been alive.

'He was a bit of a macho man. He was very jealous. And he was just five years older than me! I think he was silly really. He did not let me out on my own; only together with him. He was the type of man who believes "a woman's place is in the home", taking care of the children. Perhaps if he were alive today, now older, he might have changed a little. But at that time he would not have let me take part in the Movement as I do today.'

(Zelia, 61 years)

Zilda's relationship with her husband, however, went through a change on her involvement with the Health Movement. She says:

'At the beginning of my involvement with the Movement I used to argue with my husband a lot. He complained that I was getting too

100

involved. I even considered the possibility of splitting up. But slowly, after a lot of talking, he came to understand my involvement in the Movement. It's a question of compromise. That's the idea I try to convey to others. I find it absurd that people who are militants in movements and political parties should then forget family and children. I think that's wrong; it involves too much frustration. The ideal is to have a balanced life. The man needs the woman, and vice versa. If we achieve a balance at home, we do the same outside. I tell him I think it's good that he sometimes tries to hold me back – in the heat of the moment you can get overwhelmed and throw yourself into the work, forgetting everything else. That's when he holds me back.'

(Zilda)

The Health Movement started in 1976 when health conditions in certain areas of São Paulo, the richest city in the country, were precarious. The Eastern Zone, the poorest of five zones, was pitifully short of health centres and related facilities: whilst in the centre of the city there were twelve beds for every 1,000 inhabitants, in the Eastern Zone there was half a bed for every 1,000 inhabitants.[11] In 1980 the zone had only twenty-four health centres and needed about seventy, according to calculations done by doctors working in the area (Grupo de Educação Popular/PUC-SP 1984: 29). Health conditions in the Jardim Nordeste area were even worse, as there were *no* health centres, *no* doctors and *no* hospitals to serve about 242,000 inhabitants (ibid.: 35). The urban infrastructure, like sewerage, was also very poor. It is in this context that mobilizations around the issue of health started to emerge in the area.

In September 1976, a group of students and one of their teachers from the medical school of the University of São Paulo arrived in Jardim Nordeste with the idea of developing both clinical and political work.[12] They went to Jardim Nordeste because there was a need for doctors to organize the distribution of medicines in the area through the local parish, and they were introduced to the priest by a doctor friend. Their first objective was to open a small drugstore. When they completed that, they started attending to patients every Sunday morning. It was through the parish that the students contacted some women from the district who later played a very active role in the Health Movement in the area. They were helped by a doctor who had been working in the area for a long time and knew the local priest. The parish made possible the initial contact between the students and the local community, which eventually resulted in ties of mutual interest between the most active locals and the students. Before the students arrived, some women used to take part in the mothers' club of the local parish (discussing current problems of the country and critically debating the contents of books and magazines), where they developed their capacity to think more politically and to start challenging the more 'charitable' work (helping the poor in the slum for instance) they were doing. After meeting the medical students, the women pressed them to change their line of work from basic welfare towards assistance in organizing the neighbourhood. This gave a fundamental impetus to the beginning of the Health Movement in the district. Women and medical students started to meet to discuss health matters such as medical care, the function of hospitals, health centres, and so on. The women had it clear in their minds that there was a need for some kind of preventive medical assistance in the area, and they then decided to demand a health centre for the area from the State Department of Health, which arrived in 1977.

At this point there was the establishment of a solid organizational basis. The women decided to call the group they formed Health Commission, which is the general management body of the Health Movement. The idea was to have a name with which the population would identify the achievements of the movement. Anybody could take part in the Commission. Later, as the access of Movement participants to the local health centre was hampered by staff, the women decided to form a Health Council, a watchdog body charged with the duty of inspecting the local health centre as well as conveying demands about its functioning to the Department of Health. It was the link between the community, the health centre and the authorities. This council was locally elected, with elections in 1979, 1981, 1985 and thereafter biennially. It was officially recognized by the Department of Health in 1981 and its statutes were adopted in 1983. Both these bodies have been the backbone of the Health Movement of the Jardim Nordeste area as far as its organization is concerned, even after it joined the broader Health Movement of the Eastern Zone in 1983.

The Health Movement of the Jardim Nordeste area has now been in existence for some fifteen years (1976–91), and has been able to keep up its organization and structure and to spread its influence over a vast area of São Paulo, whereas most urban movements around consumption issues in Brazil tend to have been more short-lived, losing their strength and motivation in the face of delays or dying away once the initial demands have been met, as was the case with the Crêche Movement and the Movement Against the Rise of the Cost of Living, for instance (Evers *et al.* 1982; Gohn 1985). Unlike most of these mobilizations the Health Movement has been capable of developing and changing its demands and thus maintaining the motivation of those involved in it.

The involvement of women from Jardim Nordeste within the Catholic Church

Due to Brazil's political repression during the late 1960s and early 1970s, the Catholic Church, particularly through the mothers' clubs, provided one of the only options open to women for participation in life outside the home. The only two political parties at the time were organized on a hierarchical basis (as indeed the new political parties still are, to a large extent). They did not offer any direct channel for the people to express their views, and they neither proposed specific programmes for women nor took up women-related issues. Nor were the unions able to offer any channel for popular political expression, since they were both politically and financially controlled by the government through a number of mechanisms used by its Labour Ministry, like stringent labour legislation dating from the 1930s, state-appointed directors and manipulated elections of officers, and strict Ministry surveillance of union funds.

The involvement of women from the Jardim Nordeste area in the parish developed gradually. The common feature of their participation was the desire to be active outside their homes. On their own initiative, their work in the parish started as physical work related both to the priest and the building, such as cleaning the floors and washing his clothes. Still on their own initiative, the work then evolved to more 'relief' type of work, directed to helping the poorest inhabitants of the neighbourhood. Although the women were low income, they were better off than those living in the nearby slum.

'At the time we were devotees, always went to the mass. We started washing the priest's clothes. Then we went on to wash linen and other fabrics and ornaments of the parish, and clean the parish, all for free. From taking care of the physical things of the parish, we moved on to take care of the people who attended the parish. Especially those of the nearby slum, who we thought were the most in need. We, the good housewives, the good ladies, saintly ladies, started to work for charity.'

(Zilda)

This last phrase, said in a mocking tone, demonstrates the growing distance that the women felt from their earlier self-image as churchgoers and charity workers.

The women came up with the idea of a mothers' club, which would meet every week in one of the parish's rooms, to discuss issues involving children, women, marital relations and everyday problems. This arrangement had the acceptance and support of the priest. At the same time, the relief work continued. One aspect of the charity work entailed aiding a project of the municipal government for which they volunteered, aimed at helping those with financial difficulties. They would assist those who might lose their land for not being able to pay the instalments on its purchase, people who had opened a credit line in a shop and then could not pay for it. A social worker from the municipal government came to the local parish and contacted the mothers' club. Through it, some people were referred to social workers, who would include them in the programme. Around 1967, four progressive nuns came to Jardim Nordeste, one of whom did some work with the mothers' club. She was particularly instrumental in helping to organize meetings and debates. Up till then, in fact, the mothers' club had met only to debate everyday issues and problems. After the nun became involved, it was still the women who chose the themes, but the nun would direct the debate in such a way that it acquired a more political perspective.

'We were going through a very difficult time, when almost nothing could be said openly. And we knew hardly anything anyway. I did not know anything, nothing. Every Wednesday we met and we discussed. We chose the themes. We discussed education for children, marital problems – our own daily problems. The nun helped us to understand what was going on with our lives.'

(Zilda)

Zelia complemented this idea by stressing the contribution of the nun to the group:

'She had a different understanding of problems, so much so that she was always in conflict with the other nuns and with the priests. We discussed our rights, our roles, our freedoms. We had to look around, to see what was happening and talk more to people in our neighbourhoods. We did not discuss abortion or contraceptives.'

(Zelia)

One of the formative experiences of that period was that the nun brought books (most of them novels) and encouraged the women to read and to think about the significance of what they read. Each woman would then choose one book that she wished to read and at the following meeting each had to introduce the contents and name the author to the others. The nun also brought magazines and newspapers, and she encouraged them to listen with discrimination to the radio and television. As Zilda put it: 'She helped us to think politically.' Meanwhile, the women continued with their charity work, which in Zilda's words was their 'own initiative, one thing leading to another', mainly with the people of the slums. The ways in which the women helped the slum dwellers included organizing goods exchange, teaching the women how to sew, training in basic hygiene, and even helping to find jobs and providing them with foodstuffs. These were gathered both through the organization of *novenas*, for which all those taking part would buy a kilogram of any staple food, and through jumble sales held to raise funds for food provisions. The jumble sales also allowed the population of the neighbourhood to buy cheaper clothing. Another way of helping

104

the slum dwellers, particularly in the case of Zilda, was to give financial assistance to those needing to travel to the Tatuapé first-aid centre, about 8 kilometres away, or to the health centre of Vila Ré, which was also distant, in order to have their children inoculated or just to see the doctors.

'As my house was always a little better than the others in the neighbourhood, and people knew I was better off because my husband got a better wage than the other husbands, they would come to my house to ask for money so they could take their children to the Tatuapé first aid centre. In the neighbourhood there was no type of medical assistance; there was nothing. It was expensive for them to take their children to be inoculated.'

(Zilda)

But all this effort did not yield any great improvement in the lives of the people living in the slum. Gradually, the women who were later to start the Health Movement became aware that what they were doing was merely a palliative measure with no prospect of changing the lives of the slum dwellers in the long term. These measures were appropriate for crisis periods, but not for substantial change, as the same problems kept repeating themselves again and again.

'We could not do anything to change their situation. I was getting more and more depressed. Every time I went to the slum, I came back depressed and thinking about what I was doing. It was always tired but it led nowhere. Then one day I talked to Monica. We work hard, buy food with the money we raise, but they never have enough food and they never have enough money to take their children to the first aid centre.'

(Zilda)

These feelings were shared by other women. As Monica said, 'Neither charity nor the money raised really helped them.'

As a consequence, the women started to question the nature of the charitable work they were doing. They talked amongst themselves in order to find a more effective way of changing the situation. 'We were always asking ourselves what we could do because it wasn't our responsibility to help them in that way. They didn't need charity, they needed justice' (Zilda); 'it wasn't that we didn't want them to come to our house and ask for help. We knew that what we were doing had no effect on their lives' (Monica, 58 years).

Before this, however, the nuns had finished their period in the neighbourhood and moved out, the priest had moved on and been replaced. The women still met in the mothers' club and carried on their discussions. However, after the nuns and the priest left, a conflict

105

developed over how the club was to be run and organized. Some time later, the situation reached a climax and there was a split in the mothers' club. Some women only wanted to knit and crochet, whereas others, including the five women who later started the Health Movement of the Jardim Nordeste area, wanted to have a more political form of discussion.

'We wanted to have a more political discussion and they wanted to carry on with crochet. They did not even want to discuss the minimum wage! Some women said they didn't want to discuss this because they didn't live on a minimum wage and so they had no interest in discussing it. They realized we wanted to do political work. They had the mentality that says women should be at home taking care of the house and the children, and that political work is men's business.'

(Zilda)

Dissatisfied, the women who were keenest to take up a more political line of work decided to leave the parish, whilst those who were more interested in keeping to the traditional roles of mothers and wives stayed on. From then on, another group came into being. The mothers' club continued to meet in the parsh but only to knit, to crochet, and to work for charity. As Zilda commented: 'They were only interested in solving their own small problems; they just wanted to fill their time with leisure activities. What we wanted to do didn't concern them; it didn't affect them in any way'.

The other group, which met in the women's own houses, opted for a more political approach, aiming to find more effective ways to solve the concrete problems which affected the neighbourhood.

'Once we left the group, we carried on developing the work we wanted to do, which we could no longer do in the mothers' club. We started to think it was the state that had the responsibility to provide what was the population's right, and we had the right to demand that.'

(Zilda)

At the time (mid-1970s) women also began to take part in the Ecclesiastic Base Community groups run by the priest. Perceptions varied as to the functions and usefulness of these groups. For some, the group was seen as useful because of the opportunity it gave them to make friends with other people in their street. They would celebrate birthdays together and share the costs of the party. For others, on the other hand, it was valued for providing the conditions for discussions of a more political kind. In any case, it gave an opportunity for all to participate in an activity outside the home, which (since it was something in the parish), did not evoke domestic opposition, particularly from husbands. The participation of the women in the Popular Church helped them to articulate choices about changing the society they lived in. The women came to believe that they could and should campaign for their rights from the state, instead of doing charitable work which they now realized did not solve the problems that the neighbourhood faced. In other words, they were evolving from an individualistic to a community approach. This facilitated their political organization by allowing them to question the authoritarian structures prevailing in the country, to participate in urban movements (Mainwaring 1986), and, in the case of women in Jardim Nordeste, ultimately to question their participation in the parish itself. The contribution of the parish in Jardim Nordeste was first of all to provide a space where the women could meet each other to talk and discuss personal problems at a time when such opportunities were particularly rare. After the right-wing military *coup* of 1964, the mothers' clubs provided low-income women with one of the only contexts in which to meet without fear of political persecution. Brant (1981) states that as the pre-1964 movements were obliterated or forced into submission, a strict surveillance apparatus was put into practice by the government to avoid the upsurge of new popular organizations. In this context, the encouragement of the nuns and priests was crucial both in helping them to discuss particular issues, and in enabling women to go outside their homes to meet, talk and get away from their domestic responsibilities for a brief period. Zelia was clear: 'That was a good time for us, "nuns' women" as people called us. I think it was then that we started to go outside our homes, to learn to participate in something and to enjoy meeting and discussing'.

The parish had a role in encouraging the 'formal' equality between women and men in society particularly in relation to the responsibility for active participation in neighbourhood life (Chiriac and Padilha 1982: 198). It did not, however, discuss family structures or women's subordination to men due to their place in the division of labour along lines of gender. In other words, the Catholic Church (not even the Popular Church) did not advocate structural change in social organization, and maintained that women should still be responsible for the maintenance and care of the home. What it promoted was a call for women to be more active in neighbourhood organization so as to work alongside men for the improvement of living standards in the area, clearly considering the domestic role of women and their experience in this sphere as an important asset in that organization. However, the Popular Church also started to state that women must be respected inside as well as outside their families, and in this sense its influence on the development of the political awareness of these women was significant. In fact, some feminist ideas (discussed below) found a favourable terrain among all in the

107

Ecclesiastic Base Communities and in the mothers' clubs, which were composed of women. In order for the Church to attract these women, it had to assure them that they had the same rights as men in participating in the neighbourhood. This attitude ultimately contributed to a notion of citizenship and social equality among women (Chiriac and Padilha 1982; Corcoran-Nantes 1988), something found also in Jardim Nordeste. Zilda illustrates this very well:

> 'There is a lot to be done. There are many things that need to be changed. Everyone is responsible for things that have to be done. I think everyone should do what it is possible for her to do. Everyone should get involved in some activity in movements to develop political work.'

And Zelia completes this by stating: 'One cannot stay at home waiting. Everybody has the right and the obligation to do something concrete if the situation is to be changed.'

This has also been found by other authors, who state that most women who emerge as leaders in urban movements in Brazil have had the experience of participating in mothers' clubs of the Catholic Church and other small local organizations (Singer 1981; Caldeira 1982; Taube 1986). However, the split in the mothers' club in Jardim Nordeste and the subsequent distancing of the women from the groups of the parish was a further encouragement for them to organize without depending on the parish's support. This was a very important development as far as the community of the movement goes, in that they were capable of creating a group identity of their own.

The ideology of the feminist movement and the health movement of the Jardim Nordeste area

The Feminist Movement in Brazil is certainly the least definite of the factors influencing the emergence of the Health Movement of the Jardim Nordeste area. Indeed, it is difficult to demonstrate the precise nature of this particular influence on the mobilization of women in the Health Movement, since there is no concrete evidence which might prove a definite link between the two movements. However, this does not mean that it was not a factor which influenced the mobilization of women in demanding a health centre for Jardim Nordeste. Even though the organizations of the Feminist Movement *per se* may not have penetrated the neighbourhood under study, its ideology and formulations *had* done so (Caldeira 1987).[13] The fundamental common point was that women had the same right to participate in the neighbourhood as men had. These ideas, although generated by the Feminist Movement, were by no means confined to its adherents. On the

108

contrary, they were broadcast by the media and soon stopped being 'a property' of that movement (ibid.; Cardoso 1983b). To a certain extent, they affected even the work of the parish and clearly had a local impact, as Zelia and Zilda testify: 'We discussed our rghts, that women were not supposed to be their husbands' slaves. We are not supposed to be only at home washing. We had to go out, talk to our neighbours, exchange ideas' (Zelia); 'We discussed our lives, our problems. Problems with our husbands, with our children. We learned to think politically' (Zilda).

This is not to say that all the ideology of the Feminist Movement was discussed and accepted in the parish groups. Topics such as abortion and the use of contraceptives were not discussed at all. However, the basic idea of the Feminist Movement was equality – that women had the same rights as men – and this was very significant for the women. To understand better the influence that the Feminist Movement had on the women's decision to start mobilizing, it is important to understand a particular characteristic of this movement in Brazil. For the most part, the women who initiated feminist activities had some sympathy for left-wing positions. The struggle for the equality of women was a step towards a society where not only should there be no discrimination between the sexes, but also where there would be no allowance made for class domination. Although Latin American feminist ideas were developed by white middle-class women, they were directed towards low-income women with the intention of advancing a process of liberation through raising the consciousness of other women in another class (Cardoso 1983b). Ethnicity was not on the agenda. In the three years that followed the International Year of Women (1975), feminists broadcast their ideas whilst working side by side with low-income groups such as mothers' clubs, female union members and the Association of Housewives (in the Eastern Zone of São Paulo). Even the SOF (Family Orientation Service),[14] which later worked directly with the Health Movement of the Eastern Zone of São Paulo, had the support of feminist militants when editing their bulletins (Barroso 1982). It has to be pointed out, however, that the women in Jardim Nordeste did not see themselves as feminists and they never took part in a feminist group. They saw feminism as something distant from them. At the time of the Latin American and Caribbean women's meetings in Bertiogsa, state of São Paulo in 1985, Zilda took part. When she came back she made it clear that she had gone in order to broadcast the movement. In her opinion, other topics discussed had nothing to do directly with the movement and she felt detached from the discussions. She says:

'The movement is a Health Movement. It is not a Feminist Movement. We must have this very clearly in our minds so that we do not lose our perspective of struggle. I am not part of feminist groups because this would take my time, and I want to dedicate myself to the Health Movement.'

The significance of the spread of feminist ideology was that it became more socially acceptable for the women to organize collectively and to demand what they saw as their rights, while their participation in the parish emphasized that women should participate actively in their communities. Up till then, to demand and organize collectively was not the traditional role of the mother and wife, but this is what the women in Jardim Nordeste ended up doing, with the support of the local population.

CONCLUSION

This chapter has shown how the Catholic Church and the ideology of the Feminist Movement in Brazil influenced the formation of the Health Movement of the Jardim Nordeste area. The most visible influence was that of the Catholic Church, through its more progressive section, the Popular Church. This provided the women with a space for discussion at a time when the state's political repression severely restricted any other form of gathering and participation. Additionally, the women encountered some nuns and priests in the local parish who aided their development of a more political perspective, which in turn led them critically to question the charity work in which they had been involved. The Feminist Movement contributed in a more subtle way, by the spread of its ideology among low-income women. Church groups developed which comprised mainly women who were attracted to do the parish's works by its recently expressed support for the rights of women in the family and in the neighbourhood. The notions of egalitarianism and that women have the same rights as men were not contradicted by the parish (though it simultaneously tried to reinforce the traditional role of women in the family as mothers and wives) and it did help to raise the awareness of the women in Jardim Nordeste as to their individual rights. The influence of both the Catholic Church and the Feminist Movement on the emergence of the Health Movement of the Jardim Nordeste area was shaped by their particular evolution in Brazil.

This chapter has shown that in fact the Jardim Nordeste area was a particularly likely place for the development of a long-term, strong women's movement. There work of consciousness raising developed by the Popular Church, together with the spread of the ideology of the Feminist Movement, created in Jardim Nordeste the conditions for some women to start taking part in the Health Movement. This has been found elsewhere in Brazil. Usually, women who take part in urban movements have had a previous experience in parishes' groups. These are not the only factors affecting the emergence of the Health

Movement, but they are certainly significant. One of the others was the fact that women became aware of the situation of poverty in which they lived and how this affected the quality of life of the poor. When they gained this awareness, they accessed the state in order to demand what was their right in terms of better healthcare. The issues raised involved consumption, in the sense of collective means of consumption, in that it was the state which had to provide them.

It is not possible to say that it was one or the conjunction of two factors which was responsible for the emergence of the Health Movement. It was not just a matter of consumption. Also, it was not simply a matter of the politicization of women, or the role of the Popular Church, or even the spread of the feminist ideology. All these factors at that point in time were responsible for the emergence of the movement; but the continuity of it was the outcome of women's commitment to it. This chapter has also shown that it is not possible to neglect the issue of gender in urban movements. To homogenize social actors directly involved in movements is to underestimate the specificities of their action. These movements happen at the place of residence, and due to the division of labour along lines of gender, women are the ones who are more involved in the daily lives of their neighbourhoods. It is therefore women who will organize to guarantee the welfare of their neighbourhoods and families if they feel it has been threatened.

So, the issue of whether the movement is feminist or feminine is not central to the debate. What is central is the process of mobilization because it is within this that what changes in the lives of women and in the way gender is perceived are established.

111

5

WOMEN'S POLITICAL PARTICIPATION IN *COLONIAS POPULARES* IN GUADALAJARA, MEXICO

Nikki Craske

INTRODUCTION

Women have been central to the wave of popular organization which has taken place in Mexico in the 1980s. These mobilizations have challenged the clientelist and corporatist regime which has been dominated by one party, the Partido Revolucionario Institucional (PRI: Institutional Revolutionary Party) since 1929. This chapter looks at the role of women in two kinds of political organization, one within the cooptive structures of the PRI, the second within an independent group. The two organizations in question, the government-sponsored Federación de Colonias Populares de Jalisco (FCPJ: Federation of *Colonias Populares* of Jalisco) and the independent Organización Independiente de Colonias del Oriente (OICO: Organization of Independent Colonias of the east), are both active in Guadalajara, the urban and commercial centre for western Mexico. There are two areas where change is occurring in women's relation to politics: one is at the level of perception where ideas of self and empowerment come to the fore, and the other is at the level of institutions. My argument is that independent organizations are giving women a political experience which is profoundly affecting their lives, leading them to question the power relations which limit them at societal level, as well as within their personal, familial relations. Participation in the government's neighbourhood committees, however, does not give the same opportunities and experiences as those within the independent sector, and the women remain largely unaffected by their participation. The greatest differences between the two groups of women are knowledge of the political system and feelings of empowerment. Women in the independent OICO show greater knowledge of political institutions and the political system, and they are more confident regarding their ability to take on power positions, both personal and institutional; that is, they feel empowered by their experience. By contrast, women involved with

the PRI's FCPJ have a very limited knowledge of the politi
and they are hesitant about challenging authority. Thus, an
the FCPJ will demonstrate the clientelist and cooptive metho
PRI within the *colonias populares*, whilst OICO will be illustr
attempts to erode its support.

The importance of women in popular protest in Latin Amei ᵼs
being increasingly acknowledged (Logan 1988; Jaquette 1980, 1989;
Hess and Ferree 1987; Jelin 1990; Bruce and Kohn 1986) and studies
on Mexico are also increasing (Massolo various; Massolo and Schteingart
1987; Massolo and Díaz 1984; Lozano and Padilla 1988; CONAMUP
1988; Regalado 1986; Prieto 1987; Ramírez Saíz 1987; Cortina 1984,
for example). However, there are no studies which directly compare
the experiences of women in the independent sector with those
participating under the auspices of the PRI, as will be addressed in this
chapter. The earlier lack of interest in women's participation in much
of the literature has been due to two major reasons: politics has been
defined from a male perspective and more female-oriented groups have
therefore not been seen as political (Jaquette 1980; Pateman 1980,
1983); or their activities have not been seen at all, and therefore
not taken into account (CEPAL 1984; Massolo 1989b). Furthermore,
Mexican women were perceived as being 'naturally' conservative and
overly influenced by the Catholic Church. This ideologically constructed
account of women allowed post-Revolutionary governments to resist
pressures for women to have the vote until 1953 (Alvárez 1984). This
chapter will attempt to show how women's involvement in popular
protest is affecting the Mexican political system, both at the grassroots
as well as within the institutionalized world of party and electoral
politics, and how their participation is affecting their own perceptions
of themselves.

In much of the literature regarding women's political involvement,
especially in poor urban areas in Latin America, there is a common
view that women mobilize around what Molyneux (1985) describes as
'practical gender interests'. In other words, women have been spurred
into confronting those in authority to defend the well-being of their
families, particularly regarding material welfare. These mobilizations
have centred around the neighbourhood as opposed to the workplace.
The importance of the defence of life and the notion of 'female
consciousness' (Kaplan 1982) recur again and again in accounts. The
bedrock of female consciousness is the need to preserve life:

> by placing human need above other social and political require-
> ments and human life above property, profit and even individual
> rights, female consciousness creates a vision of a society that has
> not yet appeared. Social cohesion rises above individual rights and

quality of life over access to institutional power. Thus female consciousness has political implications.

(ibid.: 546)

The rise in women's political mobilization and its focus on reproductive issues has brought into question what we consider to be 'women's issues' and what we consider 'political'. Whilst I consider the strategic/practical gender interests divide of Molyneux a useful concept, I wish to suggest that it may not always be possible to draw such a neat distinction between the two interests; rather there is an overlap between what is practical and what is strategic and it is difficult to set clear-cut boundaries[1]. The concept of public/private in relation to women's insertion into society has also been useful when discussing female political participation, as it alerts us to some of the ways in which women's political participation has been ignored as it has taken place in the invisible world of the private (Davidoff n.d.). However, I will be using this as a continuum rather than as a dichotomy (Tiano 1984). The private is understood as domestic and interpersonal issues, and the public as institutionalized politics and formal wage labour. The use of a continuum rather than a dichotomy allows for different degrees of participation between the two extremes and the fact that many activities sit between them. Further, it indicates the unlikelihood that a woman is completely immersed in the private with no participation in the public (or that the opposite holds true for men); women have moved between the two as needs have dictated. Few women, particularly in low-income neighbourhoods, have been completely withdrawn from the world of paid labour, although their participation in the 'public' spheres of politics and work has not been constant as they respond to different pressures at different times. The 1980s in Mexico witnessed more women entering the paid-labour force as families developed new strategies to deal with the economic crisis (González de la Rocha 1988; Chant 1992). These changes bring new pressures to bear, some of which influence the women's decision to become involved in popular mobilizations.

This chapter will be structured in the following way: I will give a brief account of the Mexican political system followed by a discussion of Guadalajara as a case study. The two organizations used in the fieldwork, the PRI's FCPJ and the independent OICO, will then be examined, which leads to the substantive section on women and their involvement in political mobilizations in urban Mexico. This section will be divided in two: beginning with a discussion of the women's activities in the *colonias* and their views on their participation both inside and outside the community, followed by their attitudes on female/male relations and how they feel as women involved in politics. Finally there will be a section analyzing women as political actors before making some concluding comments.

THE MEXICAN POLITICAL SYSTEM

Mexico has long been considered a country of political stability in a region of violent regime change (Levy and Székely 1987). The period in the immediate aftermath of the Revolution (1910–20) was one of turmoil, but by 1929 the government party, the PRI, was in power and has remained so ever since. The PRI's ability to stay in power, despite several popular mobilizations against it (Knight 1990), has been due to its system of clientelism based on three confederations which organize and control civil society. These are the Confederación de Trabajadores de México (CTM: the Workers' Confederation), the Confederación Nacional de Campesinos (CNC: the Peasants Confederation), and *Une*, for 'popular organizations' which is broadly defined and includes elite business personnel, state workers' unions and residents of the *colonias populares* (low-income, frequently marginalized, neighbour-hoods), amongst others.[2] Women are most active in *Une*, which has particularly strong representation on decision-making bodies (Hellman 1983: 47–9; for the situation in Guadalajara see Sánchez Susarrey and Medina Sánchez 1987), but they are concentrated at the base away from formal power and decision-making: of the seventeen members of the National Executive Committee of *Une* there are only four women. Although there have not been the excesses of state power seen in other countries in the region, Mexican civil society has been effectively controlled through these confederations.

The year of the student demonstrations in Mexico City, 1968, has often been considered a watershed in Mexican history as it heralded a new wave of popular unrest (Foweraker and Craig 1990). The PRI's most vulnerable moment since then was in the Federal Elections of 1988, when the left's opposition candidate, Cuauhtémoc Cárdenas, caused the closest electoral contest the PRI has experienced. After much deliberation, the PRI's candidate Carlos Salinas de Gotari, was announced the winner with 50.45 per cent of the vote, and there are many who still question this result, including Cárdenas and his party, the Partido de la Revolución Democrática (PRD: Party of the Democratic Revolution), who do not recognize Salinas as the legitimate president. The issue of widely practised electoral fraud and the subsequent high rates of abstentionism has given 'politics' a negative image in Mexico, which is reflected in the women's comments on political parties below.

The post-1968 popular mobilizations have many roots. The 1960s saw a period of economic growth but it was also a time of increasing inequality of income distribution (Levy and Székely 1987), and mobilizations thus occurred in a context of increasing expectations but little in the way of material improvements for many (Jímenez 1988: 27). Urbanization also brought its pressures, and facilitated popular protest by bringing together large numbers of people suffering from the same conditions; 'needs' became demands and therefore potentially political issues (Portes 1985: 33). The PRI itself created an aperture for political mobilization with President Echeverría's reformism (1970–6), which created possibilities for popular protest outside official channels (Castells 1982: 271). This was followed by President López Portillo's (1976–82) political reforms of 1977, which legalized left-wing parties.

By the 1980s new pressures brought another burst of popular protest. The deepening of the economic crisis of 1982 created great hardship as the de la Madrid administration (1982–8) enacted austerity measures to deal with the rapidly increasing inflation and the burgeoning debt. Movements developed in defence of living standards, notably the Frente Nacional de Defensa del Salario y Contra la Austeridad y la Carestia (FNDSCAC: National Front in the Defence of Salaries and Against Austerity and Need) and the Asamblea Nacional de Obreros y Campesinos Popular (ANOCP: National Workers', Peasants' and Popular Assembly) (Prieto 1987), which established links with different independent groups such as trades unions, democratic tendencies within state unions, and other popular mobilizations. Many organized civic strikes, marches and sit-ins in the early 1980s (ibid.) in coordination with each other, whilst others acted independently. The Coordinadora Nacional del Movimiento Urbano Popular (CONAMUP: National Coordinator of Urban Popular Movements) was set up in 1981 (Ramírez Saíz 1990) to form linkages between different groups throughout the country. The 1985 earthquake which devastated Mexico City formed another catalyst for popular protest in the wake of the authorities' inadequacies in the relief effort (Massolo and Schteingart 1987). Popular protests emerged to demand an improvement in living standards, generally through service provision. However, later, particularly in the post-1985 period, the issue of the regime itself was questioned, raising issues of fair elections and accountability of representatives, which became central demands.

In all of the above, women have played a central role through their support and participation in the various campaigns (Ramírez Saíz 1987; Alonso 1988; CONAMUP 1988). The issues and demands raised by the residents of the *colonias populares*, particularly women, centre around the following: basic infrastructure (such as water, drainage, roads, electricity and street lighting); community services (such as public transport, schools and healthcare facilities); and citizenship demands for genuine representation and accountability. These citizenship demands generally do not include proposals for structural changes necessary to advance the position of women in Mexican society. Citizenship is still largely based on the ideally gender 'neutral' but substantively masculine model, and demands are kept within the narrow remit of elections. Typically, it is the issues around collective consumption which attracted the most response from both women and men, and which led to mass demonstrations and civic strikes in the post-1968 period.

THE CASE OF GUADALAJARA

Whilst it would be impossible to say that one city could be the perfect example of Mexican urban development, Guadalajara displays many of the characteristics, and consequently the problems, of urbanization typical of Mexico's major conurbations. Although its rate of growth has been higher than that of Mexico City (Uñikel 1978: 130–1), it remains much smaller, and therefore is easier to understand the city as a whole for research purposes. It is Mexico's second largest city with a population of nearly 3 million (INEGI 1991). As the commercial centre for western Mexico, its wealth comes from trade and small businesses, rather than the big corporations of its rival Monterray. A conservative city, it is very influenced by the Catholic Church, and this conservatism is reflected in the right-wing Partido Acción Nacional's (PAN: National Action Party) successes in the federal elections of 1988: the PAN won seven of the eight districts in Guadalajara. Although Guadalajara was not deeply involved in the Revolution (1910–20), it was at the centre of the reactionary Cristero Rebellion (1924–9) which drew its support from Catholics throughout the western part of the country (Knight 1986). Guadalajara is also the home of the Tecos, a fascist group based at the Autonomous University. The Tecos have good connections with local business and international organizations, and members have been active in the *colonias populares* to recruit for their anti-communist campaigns (Romero 1986; Quintana 1980).

Guadalajara was previously considered to have a good urban planning record with many agreements between the private and public sectors (Vazquéz 1989a). Consequently, many of the urban problems faced by Mexico City and Monterrey were avoided, or at least delayed. However, in the period 1940–70 when Guadalajara's population grew more quickly than any other major city and twice as fast as Mexico City (Uñikel 1978: 130–1), it became increasingly difficult for the city to meet the demands of its citizens. Although land invasions, which were common in both Mexico City and Monterrey, were seldom used, irregular settlements did develop. First they were situated in the north-eastern sector of the city, but by the mid-1980s they were dotted all around the periphery. These irregular settlements are subject to the same problems which afflict similar neighbourhoods throughout Latin America. They suffer from a shortage of basic infrastructure as installation can take several years, and community services take even longer to arrive. Understandably, these are the issues around which mobilizations start.

In my research I participated in two groups, the PRI's FCPJ and the independent OICO, which were active in neighbouring *colonias* in the north-eastern sector of the city, Lomas de Oblatos and Heliodoro Hernández Loza. These neighbourhoods are both situated on the *ejido*[3] of Tetlán and have existed for approximately ten years. They have many services, specifically electricity, drainage, water (although irregular supplies), public transport, schools and a market in Hernández Loza but they lack road paving (except on the two main roads), house numbering (which means there is no postal service), and police vigilance. Moreover, the schools are overcrowded, and there is a shortage of healthcare facilities.[4] The lack of proper paving is in itself a great problem. During the rainy season, June to September, the roads become muddy rivers, and in the dry season dust is everywhere. The lack of roads means it is difficult for some urban services such as police, rubbish collectors, gas suppliers and drinking water suppliers to get to many houses. Consequently residents, normally women, have to carry heavy loads to use these utilities. In order to gain these urban services petitions have to be made to the Citizens' Participation Office in the City Hall. The Mayor then consults with various bodies to decide on the distribution of services. It is worth noting at this point that the Director of Citizen's Participation, Dr Ricardo Zavala Hernández, is also the Secretary General of the FCPJ. Thus the leader of the PRI's neighbourhood committees in Guadalajara is also involved in the decision-making process regarding service distribution in the *colonias populares* in the city.[5]

Popular mobilizations have not been as widespread in Guadalajara as they have been in other cities, notably Mexico City and Monterrey, but they have existed over short periods (three years on average) and had some impact (Quintana 1980; Logan 1984; Regalado 1986). 1987 was the most important year for popular mobilization in Guadalajara with a mushrooming of popular mobilizations, many focusing on the issue of housing and working in concert with one another which led to gains in terms of land for housing (Regalado 1986 and personal communication). Noemí Gómez, SEDOC[6] worker and political campaigner, believes that this was due to the more open attitude of the then Governor, Enrique Alvaréz del Castillo, and the spill-over effect of successful mobilizations in Mexico City. Since then the political environment has been more hostile towards popular mobilization: it is much more difficult for popular movements to gain access to the decision makers and the mayor, Gabriel Covarrubias Ibarra, legislated that no more land was to be sold for housing until already existing *colonias* had all their basic services, a policy which adversely affects low-income families but hardly touches the middle classes. Political organizers believe this hostile environment is reinforced by the hardline governorship of Cosío Vidaurri (1988 to the present). Another problem has been the falling-off of grassroots participation partly due to disenchantment in the aftermath of the 1988 elections (Gómez: personal communication; Regalado: personal communication).

Federación de Colonias Populares de Jalisco (FCPJ)

The FCPJ is a member of the PRI's national, hierarchical structure through its membership of *Une*. Within Guadalajara the organization has two levels: the neighbourhood committees, and the general secretaries' assemblies. More women than men participate at both levels. There are active committees in about eighty to 100 of the *colonias populares* in Guadalajara, although there are at least 189 marginalized *colonias* in the city. The committees are open to all residents and participating in them does not make residents members of the PRI.[7] However, in reality the meetings are run by members of the PRI and all discourse emphasizes the benefits of the PRI and is anti all other parties, particularly the left-wing PRD. The general secretaries of the committees, plus other members of the executive, attend a weekly assembly which forms the Federation and which is affiliated to the local *Une*. According to Ricardo Zavala, the General Secretary of the FCPJ, the committees function as a forum for discussion amongst residents where they can make suggestions for improvements in the neighbourhoods. In reality, very little discussion is carried out, rather the committee is the last stage in a top-down communication channel. Residents are told of services installed in different communities 'thanks to the PRI' and informed about activities which are about to occur, especially if these will need the support of the residents, for example a public address of a PRI politician, or a demonstration to counter one from the independent sector. Regarding demand-making, the grassroots do little. Any petition is handed in to the City Hall, generally through Zavala's Citizen's Participation Office, by the general secretaries and they then wait until the service arrives, a process which can take several years.

In both Lomas de Oblatos and Hernández Loza about fifty to sixty people turn up to the FCPJ meetings which generally last one to one and a half hours. Although the participants are registered, it is easy and usual not to attend every week. The form of the meetings is unvarying. Members of the executive speak to the assembled *colonos* regarding the information they have received from the FCPJ general secretaries' assemblies. The residents engage in what I call 'passive participation', that is, they attend the meetings but do not speak out and there are few activities which they are called upon to do. The most important thing is for the grassroots to support the PRI electorally and for as many people as possible to be incorporated into the system. In keeping people loyal to the PRI, or at least stopping them from becoming converts to the opposition, the FCPJ is involved in populist measures and clientelism. These are aimed more at women, who make up 60–70 per cent of the participants, through such measures as cheap groceries and education programmes, particularly regarding childcare and nutrition. These programmes consist of a member of a relevant organization or government department coming to the *colonia* to give a talk to the residents. For those who want a political career, the committees give access to a well-organized network. The women's opinions on these benefits will be discussed below.

The general secretaries' assemblies are also held weekly. Although 60–70 per cent of the committee participants are women, which is reflected in the number of female general secretaries (Zavala: personal communication), there is only one woman on the FCPJ executive committee through her position as general secretary of the Women's Section. However, she prefers not to sit with the executive unless specifically invited to do so before each meeting (I never witnessed her sitting with the other members of the executive). These meetings are similar to those of the committees. Members of the executive speak for about an hour, and occasionally a speaker comes from a government agency to give a talk. There is an opportunity for the general secretaries to participate at the end of the meetings, but this generally consists of making specific demands or thanking 'the party' for its munificence, rather than participating in a discussion regarding strategy and long-term planning. Nevertheless, the PRI has made more of an attempt to politicize and educate the general secretaries than it has the grassroots. In the aftermath of the 1988 elections, Salinas identified the *colonias populares* as areas where the PRI had lost much support, and initiated a campaign to recoup these losses. In order to promote the PRI the FCPJ attempted to establish committees in those neighbourhoods where they did not have a base by sending general secretaries from established committees to these *colonias* to encourage the residents to set up their own committee. To do this 'education and capacitation' courses were set up to equip existing general secretaries for the promotion campaign. Although this had only just started at the time of the fieldwork, it seemed to consist only of giving a general account of the Revolution and the development of the PRI, obviously from a PRI perspective. There were also 'socio-dramas' where the general secretaries had to act out going to a *colonia* to promote the FCPJ. This resulted in a debate about whether they should admit to being from the PRI from the outset, or whether this would be a handicap. This indicates the image that political parties have generally in Mexico, and which the PRI in particular suffers from due to its high visibility and its history of manipulation throughout sixty years in power.

Regarding the structure and form of the organization, the FCPJ is quite distinct from OICO. It is a large, hierarchical organization with clearly defined separation of the leadership and base. There is little in the way of discussion where the grassroots can take part in any of the decision-making processes. Although Ricardo Zavala says that the FCPJ exists to pressurize local government for services, the grassroots have little say in this. Instead, negotiation over demands is linked to the maintenance of control and electoral support. Strategies to gain services are consequently not decided by the base, but follow a procedure established by the leadership. The decision to grant a service is based not on need, but on how best to reinforce support for the system. Thus, to reinforce a point made above, participants are passive and required only when there is a need to demonstrate support for the PRI.

Organización Independiente de Colonias del Oriente (OICO)

Independent groups have been challenging *Une* and its neighbourhood committees throughout Mexico in the 1980s through their demand-making (Ramírez Saíz 1990). These groups tend to have one of three roots: promoted by militants of a (left-wing) party; promoted by seminarists and lay people sympathetic to liberation theology; or begun by residents disenchanted by the PRI and its methods. In the case of OICO it was a mixture of disenchantment and promotion by seminarists, and later lay organizers. OICO is part of a complex structure which organizes popular mobilization predominantly in the north-east of Guadalajara. OICO has fifteen constituent groups in the two *colonias*, and in turn is a member of Intercolonias which has twenty-seven groups in six *colonias*. OICO has weekly meetings where representatives from the groups get together to discuss strategy and news, although individual groups also meet on a weekly basis as well; hence there is much opportunity for talk and discussion. All the groups are small, therefore the members know one another well and consider each other friends[8] and all feel at ease speaking out at the meetings. Not all the fifteen groups are directly involved in the political project of popular mobilization, as they also include a self-help construction group, a *caja popular*,[9] a group to organize a kindergarten, bible study, and a parents' group. With the exception of the construction group, all have at least 50 per cent female membership, while in the kindergarten and the *caja popular* the female:male ratio is higher.

The two groups involved in this study are *Grupo Promotor Independiente* (GPI: Independent Promotion Group) and *Siempre Unidos* (SU: the Always United). GPI is one of the founders of OICO and one of the most political in character. It has thirteen members, including the organizer Roberto, eight of whom are women. Its roots belong to a 1983 campaign for electricity, the supply of which was being controlled by the local neighbourhood committee. In the aftermath of its success a more cohesive group developed during 1984, and gave itself the name *Grupo*

Promotor Independiente. Since then it has campaigned around several issues such as demands for drinking water, improved public transport, and paving, and it has made formal links with the seminarists who went to the *colonias* to promote a popular education project. As the project became more established it was decided to start a group based on couples. The idea was to avoid the problems some women faced when their partners did not wish them to participate in mobilizations.[10] Thus SU was born in 1987. This has twelve members including Noemí the organizer, seven of whom are women. This group depends more on its promoter/organizers than GPI, but nevertheless engages in mobilizations and reflection on how to improve the community, generally alongside GPI.

OICO has the institutional support of SEDOC, a Jesuit organization which runs popular education projects. Part of the SEDOC project is to link learning to life experience, and hence political issues are on the agenda. The 'promoters' of OICO and Intercolonias are employed by SEDOC which gives them a great advantage over other popular mobilizations which exist in the city[11] through access to resources such as buildings, office equipment, training and expertise, and paid staff to support promotional activity. Obviously these organizers are a great asset to popular mobilizations as they dedicate much more time to dealing with the issues rather than doing this in their free time. Many of the projects are devised at SEDOC and it is here that long-term strategy and institutional linkages are discussed. OICO also enjoys close links with the parish, although the degree of support depends on the individual cleric involved; many have indeed been supportive which helps to reinforce the sense of community. Thus, parts of the Church organization act as an enabling institution with regards to popular mobilization. The women like the idea that members of the clergy support their struggle. One commented that she thought it made people take the movement more seriously. Another thought that it was important 'but that they should help more as we are struggling for justice, the same as Jesus'. She considered some clergy too frightened to involve themselves openly.

The organization and structure of OICO encourages the genuine participation and involvement of its members. The groups are small to encourage everybody to speak out and demands and strategies are discussed with everybody. Within OICO itself the relations between the base and leadership are good; however, the role of SEDOC employees as advisers and promoters is very important. They are perceived as experts and present a unified body to OICO although there may be heated discussion amongst themselves. There are four SEDOC organizers who work in the *colonias*, three of whom live in the neighbourhoods, so they are intimately acquainted with the issues. The only organizer from outside is Noemí, whose different status is compounded by being the only woman, university educated and she identifies herself as being middle class. This could indicate that it is more difficult for women, particularly from the *colonias*, to become involved at this level. The organizers set the agenda for the quarterly intercolonias meetings, which gives them a privileged position. So, whilst they try to encourage grassroots participation and questioning of decisions, there is a division between the base and promoters (Carillo 1990). The demands of OICO range from service provision to free and fair elections. Its strategies depend upon the issue in hand and upon the political moment, and include demonstrations, sit-ins, marches, commissions which negotiate with officials, workshops and education programmes, and linkages with other sympathetic organizations. Although OICO has no place in the official political structure, it is recognized as a popular organization and does gain access, if inconsistently, to officials. It has established linkages with both local and national bodies including CONAMUP. Although previously anti-party politics, it is now affiliated to the PRD, this will be discussed briefly below.

Women and political organization

In both the *priísta* FCPJ and the independent OICO the women had similar characteristics. They have low incomes as they are involved in informal sector employment, working either from home, in the local market, or in domestic service. All but two were married with children, and the FCPJ women had slightly larger families. The FCPJ women's ages ranged from 21 to 37 years, whilst for OICO it was 24 to 57 years. The most common educational level was three years of primary, but the range was greater for OICO: three have no formal education, two are trained nurses, one is a primary-school teacher and one studied to be a secretary. The FCPJ women had all had at least three years of schooling but only two had gone beyond primary school: one, the youngest, had completed secondary school, and another was a teaching assistant. All the women identified themselves as Catholics, with varying degrees of commitment, except one who is a member of the Light of the World Protestant sect, which has its headquarters in the nearby *colonia* of El Betel.

In both the organizations there are more women than men at the base. Within the FCPJ there are also more women general secretaries than men, but at executive level they disappear. OICO women take on responsibilities within the *colonia* and participate in decision making at the grassroots as much as the men. However, only one woman, Noemí, is a SEDOC organizer. So, women are important to urban, popular organizations, but their presence is concentrated at the base. In this section I will be discussing the women's attitudes to their participation at the local level, their understanding of wider political issues, and their opinions regarding being women both in the mobilizations and in society generally.

Participation in the community

As indicated above, the FCPJ women's participation consists mainly of attending weekly meetings. They express contradictory views regarding their participation: few go every week and they are sceptical about the efficacy of their participation, but they attend because they think it will help obtain services. Teresa explained that in the *colonia* where she used to live it 'was well regularized, so there wasn't much need, it had asphalt, we had light, we had water . . . the committees are stronger when there is need'. Others commented that the *colonia* lacked services because there was no unity amongst the residents and that fighting together made a difference. However, at the same time they commented on how ineffectual the system is:

'nothing but promises and no results . . . it's that we go and they talk, a lot is promised, a lot, they say "we meet on such and such a day, here we are" but when you really need them they're not around.'

(Yolanda)

The women saw infrastructure and community services not as their right, but as a form of payment for their support of the system, thus they need to attend meetings to show support. Again, they contradict themselves because they think the services will arrive anyway. Their negative attitude towards the efficacy of the committee is tempered with a respect for authority, including the general secretaries of their committees, and a belief that it is best to leave things to them. They do not believe themselves to be capable of carrying out these tasks: 'you have to know many things', but when asked what sort of things, one replied, 'they don't say such grand things, do they?'. The above comments indicate that the women do not have clearly defined material or political goals regarding their participation. They know there is a connection between participation and service provision but that the relationship is complex and one does not automatically lead to the other. Thus, there are other reasons for their participation. Three of the women expressed an interest in politics and liked to be informed, 'I'm interested in knowing about political things . . . because you don't know how to defend yourself'. The woman in question, Teresa, had received help from the committee's general secretary when there was a dispute regarding the tenure of her land. Others mentioned the benefits such as the cut-price groceries and the distribution of *tortibonos* (vouchers for *tortillas*). The women who attend most regularly are those who either do

124

not work outside the home, or who have had specific issues like Teresa. I would argue that the meeting provides these women with a legitimate reason for being involved in things outside the home as it is perceived as campaigning for the good of the family.

The women of the independent OICO were drawn to participate in the movement generally over specific issues. Those who had participated for the longest (six years at the time of the interviews), had become involved at the time of the original campaign over electricity supply. Others had become involved through the SEDOC project which resulted in the formation of Siempre Unidos, others through the seminarists, and others had been invited to join by neighbours. They all attended the meetings every week and took their share of chairing and secretarial duties. During the time of the fieldwork, few single-issue campaigns took place, although there was a demonstration to protest at the recent rise in water charges. The women said they participated in the group because they believed pressuring local government was the best way to gain services, they liked to be part of a community, and they wish to be part of an opposition force to the PRI. The youngest member, Gaby, says that although she does not think that there have been any successful campaigns during her involvement of eight months, she participates because, 'there's much support from the neighbours, you feel a lot of support. I've noticed where my mother lives there's nothing like this, you could be dying and nobody would know'. Another comments, 'on our own we can't solve problems, but united we can'. Carmen, who has participated since the electricity campaign, says that she used to be a *priísta* but through her participation she has become disillusioned with the party:

> 'Normally I used to vote for the PRI, for me the PRI was the best and since then . . . it's like you're asleep, and suddenly you wake up to the reality and you say, what's been going on? The PRI are a plague of the first . . . . When I began to realize how we struggle, how they disrespect us, how they don't listen to the protests we make, I began to question the unpleasantness of the PRI.'

Thus, by participating in OICO, they are identifying themselves as anti-PRI, as well as wanting a better standard of living. They gain mutual support from their participation and feel that together they can realize their demands, albeit slowly. They are registering their opposition to the system.

Politics outside the community

Within the context of the politics outside the *colonia*, the FCPJ women's knowledge is limited. When asked about political parties they all identified the PRI as a party, as well as the right-wing PAN, but nobody identified the left-wing PRD by name. Although three did speak of the Cardenists, they did not distinguish between the Cuauhtémoc Cárdenas' PRD, or the PFCRN which is supposedly based on the ideas of Cuauhtémoc's father, Lázaro Cárdenas (one of Mexico's most revered presidents (1934–40)). Seven of the ten women interviewed said they always voted. Two said they did not always vote, and another, Paty, had been too young at the time of the last elections. In a country where abstentionism is high these comments have to be taken with caution.[12] Most considered voting a right or a duty, and all said they had voted for the PRI. However, they were very critical of the party. Yolanda said, 'it's always the same, always the government who manages us all the time, what can we do against it? Or what can we say, it's the same as if we said nothing.' I asked her why she voted if she had such negative views she replied, 'in reality they don't leave us an option, that's the reality, because at best you could vote for another party, but they could let you sink more. As the saying goes, "better the devil you know".' Ana María thinks the PRI will win whatever, 'Well, I've heard that it always wins, that it robs votes and this and that, as I've said, it's going to win whatever, so you have to vote for it.' Teresa's support for the PRI is due to her conviction that it is the best party for her and members of her church. She belongs to a Protestant sect, the Light of the World, and believes the PRI to be less influenced by Catholicism than the PAN:

'It [the PAN] is not in our interests . . . the clergy intervene, [it's very influenced] by religion, so it is not very helpful to us. The PRI's okay because it leaves us our liberty . . . it helps us with our needs, at least here in the *colonia*, the PRI helps us a lot. It doesn't get involved with religion.'

It is difficult to assess the knowledge and actions of the women clearly and this is particularly the case for the FCPJ participants as they display a higher degree of contradiction between opinions and action. It is doubtful that they all vote, and possibly they told me they did as they identified me with the leadership, since I was introduced to the residents through the executive. However, it would not seem that they use abstentionism as a strategy as do OICO (see below). Their lack of knowledge of political parties reflects the parties' lack of success at meaning anything to these women, combined with the negative attitudes of the FCPJ leadership towards the opposition parties. They have absorbed some of the criticisms made of the PRI regarding electoral fraud but have not gone beyond this first step and do not see a way of changing these government practices. They do not understand the processes of government, such as the responsibilities of local, state and federal government, and felt that such processes are outside their own experiences.

By contrast, the OICO women named the PRI, the PAN and the PRD or the Cardenists as political parties. Their opinions are negative towards all parties, although there is a grey area around their views on the PRD. One said 'all of them are in league with the PRI' and another commented, 'I've always said that all of them promise much but do nothing', reflecting the comments of an FCPJ woman. Contrary to the FCPJ women, OICO women have abstained as a protest strategy until the 1988 elections offered what some saw as a viable alternative to the PRI, in the shape of the Frente Democrático Nacional (FDN: National Democratic Front), predecessor to the PRD. This was partly in response to the workshops that the FDN gave to inform the electorate of their existence. The PRD has since carried out more workshops, but few women have attended them, although they say they would like to. Many were hesitant about supporting and affiliating to the PRD, but others were prepared to listen to what it had to say. Carmen considered it a party more in keeping with the experiences of the marginalized communities:

> 'the PRD is an option because there are people there from our own community, from our own organization, the others don't have the mentality of struggling with oppressed people, for us the poor, probably the PRD people do have this mentality.'

These women are displaying a greater knowledge of the formal political system of elections and parties by their coherent responses than do the FCPJ women.

Female/male relationships

I asked participant women in the FCPJ and OICO about their opinion regarding women's participation in the group, and how they felt about female/male relations. The FCPJ participants believe that women attend meetings because they have more flexible timetables, but they did not have enough time to be involved with responsibilities such as leadership positions as they have a home to run. They further believe that women are more aware of problems as they deal with them as a part of women's domestic realm. As Teresa says:

> The man has his life which is work, and this is his responsibility. He comes home with the money. For the woman, the house is hers, and things such as no light and no water are part of the home, so it is her fight.

Although this is not a genuine reflection of reality, particularly for Teresa who is a *de facto* single parent whose husband lives in the United States, it expresses the division of responsibilities as they perceive them. They also express the idea that women work harder, 'one wants to fight for one's needs, and the men, no, it doesn't worry them, so they don't risk as much'. So the women see their activity in the neighbourhood as part of their duties, but that these duties should not take them outside the community. There are limits placed on them by their husbands and by the structure of the organization itself. Ana María, who attends the meetings regularly, explained 'when he [her husband] is in I can't go out anywhere, because, well, it's just that he wants me to be there, looking after him'. Another, María Refugio, commented that her husband would not allow her to be a member of the executive committee as it would mean dealing with other men. Not all women experienced these constraints from their spouses, but the above indicate the type of limitations which can be placed upon them.

By contrast women in OICO do not think that women are harder workers. On the contrary, they emphasize that men and women are equally committed to the struggle but that women are more able to fit meetings and demonstrations in with their other responsibilities. However, they do think that women are more enthusiastic. Angelina comments:

'women take more risks, they're more enthusiastic about doing things, the men are the ones who say whether it's okay or not, but I've seen how most of the ideas come from women, we suggest this, or what do they think about that, and they approve or disapprove.'

It is interesting to note that women from both the FCPJ and OICO mention that women are more prepared than men to take risks. This idea is further reinforced by SEDOC organizer Noemí, who believes that women are more enthusiastic and become involved quicker but that they tend to lose interest more rapidly as well (personal communication). Another comments that the men are more involved in strategy-making rather than in the action itself: 'they tend to say how we're going to deal with this . . . but it's the women who are the ones who actually go and confront the authorities'. These two quotations demonstrate how important women are to mobilizations but that there still exists an implicit idea that men have more say in the decision making, although my own observations were that ideas were discussed by all present. However, men still often emerge as leaders even in women-dominated mobilizations (cf. Mohr Peterson 1990: 21).

The problem of husbands constraining their wives' participation is also a problem for independent groups such as OICO. Of the fourteen women who participate in SU and GPI, three have problems with their husbands regarding their participation. Another has problems which affect her work life rather than her activities in the group. One woman, Guille, has to be very careful not arrive home late, even if accompanied, as her husband does not like it. This is quite a problem as the meetings always start late and go on longer than intended, finishing at eleven o'clock at night, or later. Another two have husbands who are alcoholics which places extra strain upon the women. Both have taken rest periods from the group, but to date have always returned. Another one commented that, although her husband has no objections to her participation in the group and is a member himself, he does not like her to work outside the home. She comments that he believes '[she] has the duty to wait on him'. After changing her career from secretary to dressmaker, she now works from home.

The above examples cited by women, both in the government-sponsored FCPJ as well as the independent OICO, show that women from both groups have to deal with issues concerning husbands' authority. However, it is not only the husbands who place these constraints. Carmen's spouse is living and working in the United States, but she is still constrained by her children's attitudes:

> 'when they see I'm off to a demonstration or something, they say "don't even mention that we're your sons" they don't like it ... they say "you're like one of those gossipy old women, you stay at home"'.

Further limitations are placed on them by social norms which dictate the hour at which women can arrive home at night, with whom they go out and under what circumstances participation in meetings is legitimate. Activity which occurs outside the *colonia* is particularly problematic as buses are slow and irregular. Consequently, getting to meetings is time-consuming, arriving back at the prescribed hour is difficult, and women cannot always go accompanied. Further, many feel they should be at home for the children when they return from school.

However, for the main part, participation in the PRI's neighbourhood committees does not require this type of activism. The FCPJ does organize events frequently aimed at women, but within the neighbour-hood. I witnessed talks by nutritionists and childcare experts which appear to reinforce the women's sense of being unknowledgeable and/or inadequate, compounded by the fact that there is little interaction between the speaker and the participants, as the latters' personal experiences are not seen as giving the grounding necessary to partici-pate in such a way. It could be said that these talks open up a space for women within the structure, but this is limited as the women themselves do not define that space.

The women express further views regarding the position of women in Mexican society when discussing the future of their children. They want their daughters to marry later and have greater access to educa-tion. One commented that by marrying young she had not had the opportunity to get to know other women. They wished that they themselves had had smaller families to give their children greater opportunities. They felt that men should take a greater share in the domestic work, although they accepted that it was predominantly their responsibility. In making these comments the women are expressing a critique of their own experiences and where they think there should be changes to make female/male relationships more equitable.

The women in the independent OICO are less constrained by social norms as the very fact that they are involved in such groups is outside 'normal' behaviour in Mexico for both women and men. There is an attempt by the group generally to ensure that problems such as arriving home late, travelling long distances to be involved in an activity or any other potentially problematic incident for women are dealt with through collective responsibility. Nevertheless, the views expressed by Gaby and Guille's husbands do demonstrate a pervasive attitude towards women in Mexican society. The structural problems faced by the OICO women are different from those encountered by their FCPJ counterparts due to the nature of the organization itself. There are no activities which are aimed specifically at women which could thereby marginalize women from the mainstream. However, the lack of women in promoters' roles indicates the extent of the limitations on women and must be acknowledged as a potential problem, especially if this remains the case in the long term. Nevertheless, Noemí's role of promoter was commented on by some women as an example of what women can achieve and how they are as capable as men.

The issue of *machismo* and how men exercise and understand their authority was discussed by the women when thinking about female/male relations. This was particularly the case for the women of the independent sector who were more inclined to question men's attitudes towards women. Gaby comments on the way her brother-in-law treats her sister, 'I tell myself he's my *compadre* and that I should respect him, but at times I can hardly bear it, he's so *machista*.' She comments further that women are as capable as men but they have been disadvantaged as they have been considered subordinate to men. Women from both groups know that women are treated differently from men and that this disadvantages them, but they are only just beginning to question this and are unsure how to raise it as a problem.[13]

WOMEN AS POLITICAL ACTORS

The case studies discussed above reinforce much of the literature about women's participation in political mobilizations in Latin America: they are active at the grassroots, in community and neighbourhood organizations, but not all participation affects their perceptions and understanding about politics, female subordination, and unequal power relations at the personal level. Perelli (1989) suggests, for example, th women in Uruguay played an essential part in the downfall of th dictatorship, but that they themselves remained largely unaffected by the experience. This is mirrored by those women participating in the *priísta* FCPJ. Even at General Secretary level, women repeat the official rhetoric heard at the FCPJ assemblies.

Women are an 'issue' addressed by the current President, Carlos Salinas de Gotari (1988–94), who has updated his rhetoric; however, these changes amount to little more than rhetorical window dressing. Women are being spoken of as an important source of support for the PRI (Salinas 1989) but real changes are harder to detect. In the internal party reorganization which led to the formation of *Une*, women were singled out along with 'youth' as areas for special attention. In *Une*'s document 'Political Programme: Action Strategy Directives' section F is devoted entirely to women, and they are mentioned as a separate constituency in many other sections (along with youth). Strategies suggested include: '(A) women's programme should be an obligatory characteristic of all movements within *Une*'; that women 'are an integral part of society and its challenges'; and that their full participation is necessary for national development. Further, women should be represented at all levels of the organization. Salinas nominated a young woman, Silvia Hernández,[14] as General Secretary of *Une* which was later ratified by the National Assembly. However, there seems to be little clear idea about how women are going to be more equally represented within the organization. There is talk of having 30 per cent female representatives at all levels, but quota representation does not necessarily mean that the most important issues regarding women will be addressed. Some strategic gender interests, such as structural changes within the organization to enable more women to have a say in the decision-making process, have found a voice in *Une* through the middle-class, university educated women who may have been influenced by feminism. However, the women in the *colonias* do not even know that a reorganization has taken place, and even less what the programme for women is. Nevertheless, with time these ideas could become more generally accepted and the support of the elite is important in a country where *caudillismo* (strongman [sic] politics) is the dominant discourse. Further, the changes in PRI discourse reflect women's increasing visibility and potential both in formal and informal politics,[15] and indicate how the concept of politics is widening, hence the increasing importance of neighbourhood organizations within the PRI's system. But, whilst this may be positive, women's concerns are still addressed only if they will win votes. Similar situations have arisen in other Latin American countries, as Molyneux observes in Nicaragua: 'The program for women's emancipation remains one conceived in terms of how functional it is for achieving the wider goals of the state' (1985: 251).

The PRI is not only identifying women in its search for more support in the electorate, it is targeting *colonias populares* generally, which is indirectly addressing women due to their high profile in neighbourhood organizations. The FCPJ is an organ of the PRI and its regime, and as such does not encourage politicization amongst the participants, be they women or men. Nevertheless, the PRI is changing its discourse on many issues, many of them raised by the increase in grassroots political mobilizations. It is redefining its corporatist project through the PRONASOL project which administers, amongst other things, services to the *colonias populares*. It is also appropriating the notion of 'citizenship' and defining it in such a way that good citizenship means participating in the neighbourhood committees, which in turn means quicker service provision. Thus, if the services do not arrive, it is not a problem of bad government but the fault of bad citizens (Peterson 1991). This reinforces the idea that services are not a right but a privilege.

Within the independent sector, women's participation has also had its impact. Women are acknowledged by participants, activists and academics as being the base of these movements. This has been reflected in organizations like CONAMUP forming its women's encounter groups (CONAMUP 1988; Stephen 1989) which are beginning to question power relations at all levels. Women are increasingly aware of the difficulties and injustices they face as women and wish to address these issues but are nervous of prioritizing them, particularly where this leads them into confrontation with friends and spouses. They identify their principal oppression as being poor, and thus in solidarity with men. Within OICO these issues have yet to be clearly addressed, as the women's group mentioned earlier remained at the planning stage. The initial discussion about such a women's group illustrated the difficulties the group faces in deciding an agenda, partly due to the group's Catholic basis limiting discussion of issues such as fertility control and divorce, and in part because they do not wish the group to come into direct confrontation with male members of OICO. (The development of an agenda for women's groups is discussed in Stephen 1989.)

The problem of deciding upon an agenda also highlights the difficulties in deciding what are practical and strategic gender issues. Issues such as the lack of access to childcare, divorce and fertility control fit into both categories. These are both practical day-to-day problems which women have to face, as well as being central constraints which limit and control women. Thus we have to be aware of the overlap and lack of clear-cut boundaries between the two categories. I suggest it would be preferable to follow Tiano's example with the public/ private dichotomy, and transform the gender interests dichotomy into a continuum which would allow us to use the concepts in a mutable and unfixed way.

The pervasiveness of the female/male divide is indicated further by the activities within the OICO carried out by women and men. A sexual division of labour operates within the organization, particularly evident at fund-raising and promotion events held by OICO in the *colonias*. The women concentrated on providing the food whilst men dealt with the 'technical' things such as sound equipment. Whilst women were able to display their private skills, such as cooking, in a public environment at these events, there was no suggestion that the women might wish for a day off from this domestic work.

OICO women are affected by their participation. A sense of empowerment and confidence results from taking part in things they had no access to before. Although confronting officials was nerve wracking, Gaby says, 'I now feel more confident in myself.' Also the women felt they were learning to be more critical, instead of believing everything on the television and the radio: 'now I realise what things are really like'. Carmen comments that people tell her that they learn things from her, and this made her realize how much she had changed:

> 'the Carmen of four years ago who didn't go out, who spent all of her time at home, who didn't want to know about anything outside Carmen, who suddenly wanted to monopolize everything. . . . I've grown in this respect, I'm no longer afraid to speak with the governor . . . and at times, just before going in, I don't even know what I'm going to say, but more than enough words come out, and four years ago, when did Carmen get involved with this?'

Not mirrored by the women in the official FCPJ, this confidence affects the OIOC women's attitude towards the political system, towards women's place in the community and towards their spouses. These changes are quite subtle: Gaby may now work from home but she insists that her husband takes a greater role in childcaring responsibilities; Guille continues to go to the meetings although her husband objects; the small size of the OICO women's families in comparison to those of the FCPJ indicates that they might not discuss fertility control but they practise it. Organiser Noemí says that these changes are small but significant as women are learning to assert themselves more.

CONCLUSIONS

Mexican women have become visible through their participation in the popular mobilizations of the past twenty years. The increase in urban politics is largely due to them; however, they are still the backbone of the movement rather than its head (Stephen 1989). SEDOC organizer

Noemí Gómez believes that the political model is a masculine one which does not necessarily fit in with the women's own experiences, be they from the independent or government sector. Women's political involvement can often seem to be contradictory as they respond to different pressures placed upon them as women, mothers, wives and members of low-income settlements. Nevertheless, Gómez believes that without the women there would be no popular mobilization in Mexico today. Massolo agrees, saying that urban politics are women's politics (Massolo 1989a: 1). Certainly all the groups in Guadalajara that I knew personally, or knew about, enjoyed greater support from women. At a personal level the women of the independent sector are growing in confidence. They are learning to make demands on the system outside of service provision and they are becoming empowered through their participation (Logan 1988). Whilst the popular organizations themselves are creating spaces in which to mobilize, the PRI is also creating spaces for civil society to act. The shifts in power created as the PRI renegotiates its position, both through internal reorganization and *vis-à-vis* civil society, allows apertures where both political mobilizations in general, and women in particular, can make and win demands.

At the institutional level there have also been changes. The PRI has taken on board some of the issues brought to the fore by women and popular mobilizations. Whether the 1980s mobilization will lead to the PRI eventually ceding power is uncertain, and if it can right the economy it could maintain enough support to remain in power. The opposition PRD has been able to establish itself as a viable electoral force, with much of its support coming from the women-dominated popular mobilizations. However, the anti-party attitude of many such mobilizations and the women themselves means that institutional links may be difficult to establish. OICO is currently in a period of policy change regarding supporting political parties, thus the women reflect this change in their own opinions. The decision to affiliate was a controversial one not accepted by all members of the grassroots who feel it is premature.[16] Whilst the PRD and popular struggles generally have made their mark, it must be remembered that most of the independent mobilizations have developed in an ad hoc and flexible way which has allowed them to respond to different situations quickly. However, turning them into long-term institutional opposition movements may be problematic, especially in a climate where political parties and electoral politics are generally viewed negatively. Nevertheless, these movements have made their mark and women have been very visible and active within them, and the success of popular protest in Mexico owes a lot to the strength of women participants. Consequently, women are respected as political actors by the two main parties, the PRI and the PRD, as well as by independent political organizers. Further, the neighbourhood,

134

which has been identified as community and domestic and therefore part of women's private sphere (Logan 1989: 174), is now being seen as an arena of political struggle. There has been a shift along the public/ private continuum making some issues and actors more visible. Thus, whilst individual women may move in and out of political activity as dictated by circumstances, women in general have established themselves on the political map more on their own terms than previously.

6

FEMALE CONSCIOUSNESS OR FEMINIST CONSCIOUSNESS?

Women's consciousness raising in community-based struggles in Brazil[1]

Yvonne Corcoran-Nantes

Generally politics is thought of as a man's world, a place where women rarely appear. In Latin America in particular, it has been easy for political science to ignore women in institutional politics such as political parties and trade unions because their level of participation is low. In the 1970s, political research on the question of political participation focused on institutional politics in spite of the fact that democratic governments were not a prevalent feature of Latin American politics. Consequently, this focus produced a very superficial picture of the political participation of both women and men (Jaquette 1980). Blachman (1973), in a piece of political research on women, unique of its time, concluded that Brazilian women believed that politics was for men not women. What he failed to indicate was what kind of politics women were speaking about. In Brazil it was precisely in that period when the majority of the population were excluded from institutional politics that non-institutional politics, in the form of popular social movements, became the principal arena for political participation and opposition (Kucinski 1982).

Later research has shown that women played a significant role in popular movements throughout that decade (Blay 1980; Tabak 1983). There is now evidence that similar levels of participation existed in many other Latin American countries including Peru, Mexico, Chile and Ecuador (see Chuchryk 1989; Moser 1987; Brydon and Chant 1989; Evers *et al.* 1982). Yet recent work on the nature and extent of non-institutional politics in the Latin American political system has failed to address the question of gender in this sphere.

In the last ten years the politics of feminist research in this area has been on the one hand to expose the suppression of historical evidence about women's role in political action for social change (Carroll 1989; Kaplan 1982; Thomis and Grimmett 1982; Tilly 1981) and on the other to dispel some of the myths about the nature and extent of women's political participation in contemporary society (Siltanen and Stanworth 1984; Randall 1987; Bourque and Grossholtz 1986). The importance of this body of work for the expansion of research into the specificity of women's political participation and role in social and political change in a given society is immeasurable. At the same time there has been a growing volume of work on women's participation in grassroots politics from the role of peasant women in the Russian Revolution to the political impact of women's protest at Greenham Common (Clements 1982; El Saadawi 1985). The disclosure of the scope of women's political participation has led to the conclusion that 'we can speculate, if not yet prove, that women's direct action has played a major role in every significant movement for social change and liberation in the 20th century' (Carroll 1989: 20). This raises the question of where women's interests lie in these movements and whether they have been able to pursue what Molyneux (1985) has termed strategic gender interests through their political action.

My own research in Brazil, from which the empirical data in this chapter originates, was designed specifically to look at aspects of women's political participation in popular urban social movements in order to provide explanations for women's high profile in non-institutional politics. It was clear that by the 1980s women participated in and led popular protest around a wide variety of issues related to urbanization, employment and the provision of basic services which suggests that this political arena represents something of fundamental importance to women (see Safa 1990; Corcoran-Nantes 1990; Moser 1987). Moreover, the issues that interest women represent the major social and economic problems in developing countries and are also becoming key issues in the advanced industrial nations.

Women of the working poor in Brazil have, over the past two decades, strengthened their presence in non-institutional politics by protesting about the lack of basic services, health provision, transport, housing and unemployment. The methods of organization and political practice of popular urban social movements demonstrate the influence of women in them and similar practices can be found in many developing countries both in grassroots protest politics and women's organizations (see, for example, Mies 1988; Mattelart 1980; Cutrufelli 1983). I have argued elsewhere that these movements were not specifically created *for* women. It is women who form the majority within a social group whose socio-economic experience in Brazilian society is neither reflected nor represented in other forms of political organization (see Corcoran-Nantes 1990). Consequently, women have played a major role in the formation and development of popular movements. Through their participation they have discovered a new public identity in a political sphere which, in many ways, they have made their own. Many women, through their political development in non-institutional politics, have gone on to extend their participation to political parties and trade unions as well as strategic gender protests along with women's organizations and feminist groups around issues such as birth control, rape and domestic violence.

What seems to have developed is a bi-furcated political sphere: male/institutional politics and female/non-institutional politics which are identifiable by the nature of political/gender organization and action. The political practice in either sphere bears little resemblance to the other as in one the majority of political actors are male and in the other female. What I wish to consider here is the gender specificity of political practices in non-institutional politics and their implications for the relationship between politically reproduced gender spheres.

This chapter will look at the development of political consciousness and solidarity among women of the popular urban movements in São Paulo. It is through these practices and women's influence on them that we can analyse the motivation behind their participation in this political sphere, how women view their role in society and what this represents for them. By looking at the various processes involved in conscientization and politicization such as forms of consciousness raising, self-help groups, oral history and the struggle for literacy I will argue that the development of women's political consciousness is far more complex than present analyses demonstrate. In Brazil, as in many other Latin American countries, women have created a political role for themselves based on their social status as wives and mothers but through which they have struggled for recognition of their roles and rights as workers, residents and citizens.

GENDER: THE MISSING LINK IN ANALYSES OF POPULAR SOCIAL MOVEMENTS

Despite evidence to show the predominance of women in non-institutional politics, those who have attempted to analyse popular social movements have tended to ignore the question of gender. Those who have acknowledged or tried to give some explanation for women's participation and political consciousness tend to fall into two different camps. First, there are those like Jaquette (1989) who subsume female participation under the auspices of feminism or women's movements thereby removing the question of gender from their analysis of popular protest in Latin America. Second, there are those like Chaney (1979) who prefer to confine themselves to a matrifocal analysis where the traditional role of women as wives and mothers and their relation to the reproduction of the labour force becomes the universal explanation, the *sine qua non,* for women's political practice and participation. The role of women in Latin American society is far too multifaceted for us to be satisfied with unitary explanations of their participation in the political sphere.

It was in 1985 that David Slater's edited collection on 'new' social movements in Latin America set the agenda for the analysis of non-institutional politics. The proposed aim of this book was to consider the political conditions which favoured the development of 'new' social movements, to identify the specific characteristics which illustrate what is 'new' about them and to consider their role in class struggle in the region. Within the book the entire gamut of social movements are touched on, from urban guerillas to feminists. The major flaw of the book is that it completely ignores the question of gender as one of the keys to understanding the political nature and practice of these move-ments[2] – all the more surprising when two of the contributions to the book are written about social movements in São Paulo, Brazil, where my own research was based. Women's political participation is relegated to a separate section at the end of the book under the auspices of 'women's movements'. Curiously this marginalization of women in a book in which 'malestream' political analysis predominates, accurately reflects the political division of labour which exists in political parties and trade unions in Latin America and many other parts of the world. Within institutional politics women's groups and women's departments are formed to present programmes and drafts of new laws or to develop a strategy to place women's issues on the political agenda. What this actually does is to take strategic gender issues and other issues that are of importance to women out of the mainstream politics within these organizations. In short, it removes them from the political agenda. By placing the question of gender in a separate section of the book, it is removed from the main body of the discussion. Consequently, gender, instead of constituting a central part of the political analysis of social movements, becomes peripheral to it. By an extremely deft sleight of hand the onus for incorporating gender into the main political analysis is removed.

The other problem created by the structure of Slater's book is that by tying women's political participation exclusively to the question of feminism and women's movements this gives a wholly inaccurate picture of the extent of women's political participation and the motives and interests behind it. The majority of women who participate in the popular social movements are not motivated by a feminist consciousness; feminism for them has very little to do with the reality of their lives. In Latin American society the marked inequality in the distribution of wealth and resources has further reinforced the idea, among women of the urban poor, that feminism is a middle-class ideology for women who

ave all the social and economic advantages. Moreover, the institu-
ionalization of domestic service on the continent has sustained the
antagonism between classes whereby 'fortunate' women exploit 'less
fortunate' women. The patron/client relationship which has evolved is
a major barrier to any longstanding political association between them
and there have been occasions when this relationship has been politically exploited (Corcoran-Nantes 1988; Filet-Abreu de Souza 1980;
Chuchryk 1989). Consequently, there are considerable class differences
in relation to how and in what forms of political organizations women
participate.

Writers who do analyse women's participation in popular social
movements tend to consider only one aspect (that of the sexual division
of labour) as an explanation for their participation, either relating this
exclusively to women's domestic role or this together with women's
relationship to collective consumption (see Cardoso 1984; Safa 1990;
Moser 1987; Evers *et al.* 1982). I have argued elsewhere that women's
participation in social movements is also linked to their role in production and that most of the issues around which these movements are
organized also affect men (Corcoran-Nantes 1990). Without doubt
women do legitimate their entry into the political sphere as wives and
mothers but there are tangible reasons why women do so. In Latin
American society *marianismo*, or the cult of Mary, still exists whereby
women's status in Latin American society comes from their reproductive
role and this has often been a source of power for women (Stevens
1973). By utilizing this image women can strengthen and legitimate their
political involvement in the eyes of the state. Conversely the state has
also exploited the cultural identity of women, *os supermadres*, to secure
their political support (Chuchryk 1989). The *supermadre* approach to
politics is legitimated by women, men and the state. Up to the present
time no one has considered, for example, how far men take their role
in the family into the political sphere.

Maxine Molyneux, the exception that proves the rule in Slater (1985),
in her own work on women in post-revolutionary societies has presented
an excellent working hypothesis for considering the motivations and
achievements of women's participation in political struggles. She divides
gender interests into two broad categories. One is strategic gender
interests which are directly related to women's subordination in a given
society and the demands around which women's struggles are based on
a strategy to overcome all forms of gender inequality. The other is
practical gender interests which derive from women's ascribed role in
the sexual division of labour, a response to their immediate practical
needs and formulated by women themselves. These are shaped by class
and ethnicity and are not necessarily part of a long-term strategy to
achieve gender equality. She goes on to argue that in the formulation

of strategic gender interests practical gender interests have to be taken into account and it is 'the politicisation of these practical interests and their transformation into strategic interests which constitutes a central aspect of feminist practice' (Molyneux 1985: 236–7).

It is practical gender interests which are the basis of women's political participation in popular social movements. The transformation of practical gender interests into strategic gender interests requires not only women's recognition of their power to represent their own interests but also that space exists within the prevailing political system to pressure the state into recognizing those interests. This is part of a complex political development process whereby women not only recognize gender interests but do so in relation to and in conjunction with other women, across class and ethnic boundaries.

POLITICAL PRACTICE – THE DEVELOPMENT OF POLITICAL CONSCIOUSNESS AND SOLIDARITY

In Brazil the political *Abertura* (opening) of the 1970s stimulated the development of opposition forces outside institutional politics. By the 1980s the number and type of social movements multiplied as popular protest increased and state governments were constantly pressurized into taking action to ameliorate deteriorating socio-economic conditions amongst the majority of the population. In São Paulo, hardly a week went by without some form of political protest taking place and it was here that the sense of exclusion from the benefits of highly concentrated economic growth was most acute. São Paulo was the hub of Brazilian industrial growth and with a migrant population growing by over 50,000 per month this exacerbated existing problems associated with economic development (Censo Demographico do Brasil 1990). In the urban periphery the *favelas* (shantytowns) expanded rapidly and this placed enormous pressure on the already limited capacity of public services such as water, electricity, sanitation, transport, housing and healthcare. The state of São Paulo was both unprepared and ill-equipped for such an influx of workers, having neither the infrastructure nor the resources to cater for the needs of its rapidly rising population. However, it is impossible to ignore the state government's neglect of public services during one of the most prosperous eras of Brazilian history, the period of the 'economic miracle', when it was undertaking some of the most ostentatious building projects in the capital of São Paulo. Nowhere were the socio-economic consequences of the government's political priorities more apparent than on the urban periphery, where the majority of workers lived and where most incoming migrants were forced to settle. The residential communities on the urban periphery of São Paulo were in turn, not surprisingly, the main political bases of the urban social movements (see Corcoran-Nantes 1990; Machado 1988).

It is on the urban periphery of São Paulo that the fieldwork on which this chapter is based was carried out and concentrated on three popular movements: O Movimento de Favela (The Favela Movement – founded in 1976 to secure land title and to improve services and infrastructure in the settlements). O Movimento de Saude (The Health Movement – formed originally in 1973 with the aim of improving medical services at both the local and regional level) and O Movimento dos Desempregados (The Unemployed Movement – formed in 1983 to solve the immediate problems of unemployment through demands for unemployment benefit, funding to set up worker cooperatives and so on).[3] These popular movements are representative of the wide range of movements which existed, and continue to exist, in the urban periphery of São Paulo. One common feature is that participants in these movements are either exclusively or predominantly women.

Health, housing and unemployment are typical of the kind of social questions that have attracted the interest of low-income women, and movements formed around these issues developed characteristics which reflect their involvement in them. The key factor to take into account here is that while the sexual division of labour tends in general terms to confine women for a large amount of their time to their homes and immediate neighbourhoods, low-income women are often involved in a wide range of activities which span the rather arbitrary divide between production and reproduction, with many having various modes of generating income (see Brydon and Chant 1989: 10–12). Women thus tend to be more responsive than men to issues that relate to socio-economic activities in *both* the public *and* private spheres. The political participation of women arises from the social bonds which are created via these activities in the community, through which they organize themselves and from which the political contexts of urban social movements are developed.

Moreover, the social, economic and even moral issues which have formed the basis of this type of political protest are directly associated with the nature of dependent capitalist development. They are issues which cannot be solved in the short term and are precisely those which have remained on the periphery of 'mainstream' politics or have been given little or no priority in the programmes of parties and successive governments. It is hardly surprising, therefore, that some social movements have been in existence for over two decades and have only gradually gained improvements in the conditions of life for the urban poor.

The longevity of the three popular urban social movements was due to the political practices and activities utilized by them with the flexible parameters of the political space in which they operated. Their political practices derived from the collective interests and needs of those who participated in them as well as the kind of political action required to sustain the movements over a long period of time. Each movement had its own methods of developing political consciousness and solidarity but the purpose of its activities and practices was fundamentally the same – to reinforce the relationship between the political activities of the movement and the social and economic activities of the participants. The women in these movements found it difficult to undertake a political role unrelated to their socio-economic experience in society and the kind of issues which they considered to be politically important. The inter-mingling of social, economic and political activities which characterized the political practice of the popular movements was attributable to the participation of women.

The forms of conscientization practised in the popular movements had both an educative and political function and covered a wide range of activities. The political practice of these movements was a means by which they could, from their inception, build a multilateral system of relationships and reinforce their political base. The need to develop a sense of commitment and solidarity amongst the participants in the movements was paramount to their political survival.

In each case the means by which this was done developed from their collective experience of political organization and political action. The activities and practices undertaken by the popular movements were, therefore, a response to a consensus of opinion on how each movement should organize itself and mobilize the popular classes. In the absence of consensus, many types of political practice were either discarded or abandoned. This was usually the case when they had been imposed on the women and men in a particular movement by either 'external agents' or ideological groups within the movement itself (Corcoran-Nantes 1988; Wanderley 1980). The kinds of political practice to be discussed here are those which have been used successfully by the popular movements over a considerable length of time and which have served the social and political needs of the popular classes for collective action and organization. These are: consciousness raising, educational schemes, fund raising activities, cooperative work and mutualist schemes.

Various forms of consciousness raising were used by the popular movements and the ways in which they were implemented differed from one movement to another. First, there was instruction in socialist theories to explain the socio-economic conditions of the urban poor and the importance of the popular movements to the struggle for political change in Brazilian society. This almost always involved the help of supporters of the movements such as the Church or political parties who had political material designed for the conscientization of the urban poor. Second, there was a form of consciousness raising which was a means of self-education and collective counselling and dealt with issues and problems arising from, and related to, their struggles. This was a method of consciousness raising which women preferred to use in which self-help and the dissemination of information was a part of the process of political participation. The question of self-help also led to the formation of cooperative schemes in low-income neighbourhoods to provide practical solutions to the immediate needs of the local community. Third was the use of oral history which was probably *the* most important and effective way of creating political consciousness and solidarity. All the popular movements had some means of recording and registering their political history but it was only in those movements in which women had organizational control that a strong emphasis was placed on the use of oral history as a means of conscientization.

Class consciousness *vis à vis* female consciousness

Consciousness raising based on the propagation of socialist theory was undertaken in various ways. In many cases it was an integral part of the political meetings of the movement itself during which an individual or group, usually with experience in political parties or left-wing Church groups, discussed current political issues relating to the problems of the urban poor. Sometimes it was little more than speech making but on other occasions simple visual aids such as diagrams or cartoon pictures were used to show the participants how and why they suffered in Brazilian society. Women were quick to point out that in much of this material they were under-represented and *their* experience in Brazilian society was rarely discussed at all. More sophisticated material was sometimes used, such as films or slides borrowed from the Catholic Church which has produced entire courses for the politicization of the popular classes. However, films or slides were usually shown at separate meetings because of the amount of time needed for presentation and open discussion.

These forms of consciousness raising stimulated interest and discussion both inside and outside the meetings but this was difficult to sustain over a long period of time. Women, in particular, lost interest fairly quickly because their political concerns centred on practical difficulties rather than theorizing struggles. It was, however, a successful way of selecting potential political activists. Experienced activists from political parties or left-wing ecclesiastical groups who participated in the popular movements would often utilize this form of consciousness raising to 'recruit' new activists by inviting those people who demonstrated some 'political aptitude' to attend their meetings. There were a significant number of women who entered local party politics in this way. The conflicting views of two women demonstrate the positive and negative aspects of this kind of consciousness raising:

'I liked the film best. It showed how we live and how we suffer. It made us think about why we are poor and always will be the way things are. But how are we going to change the world when it's all we can do to get a few changes in this one *and* after so many struggles . . . ? If we did change society would it be any different? I doubt it! It'd be alright for the men, it's always good for them. But what about us women? Nothing changes for us. It could be worse. . . . It's funny they didn't say much about our suffering in the film.'

(Reginalda, 32 years, São Matheus)

'Most people don't like to talk about politics, some of them leave the meeting but I think it's interesting. The pyramid was good, where they showed the different classes in different colours I think everyone liked that. . . . It was after that *favela* meeting that Manuel gave me a book to read about women and struggle and invited me to a meeting of the PT (Workers Party) group. I go to the meetings every week now. I think we women should participate in political parties, not just the popular movements, because we can learn a lot more but most of the women in the *favela* aren't interested in political parties. . . . I don't know why.'

(Valdiva, 23 years, Vila Sezamo)

This form of consciousness raising, although less successful at local meetings of the popular movements, tended to be used far more in the Encounters and Congresses held by the movements every year. Films and slides would be shown, speakers from the Catholic Church, political parties and the trade unions would be invited to discuss the role of the popular movements in class struggle in Brazil and debating groups would be set up to discuss the issues raised. The local representatives and coordinators who attended them were far more receptive to this form of consciousness raising having had greater contact with 'mainstream' political organizations than the rest of the participants in the popular movements. Many of the organizers of the popular movements had, in fact, participated in formal political organizations and were therefore familiar with this form of politicization. Thus, consciousness raising which treated the ideological issues of political participation tended to be used. Nevertheless this form of consciousness raising in the local community was rarely used on a regular basis and in some cases the practice was abandoned altogether.

Female consciousness, women's groups and self-help

Consciousness raising practices which dealt with issues and problems arising from the struggles of the popular movements were extremely popular amongst women participants. They invariably took place at separate small group meetings, often in people's homes. Meetings were extremely informal and were not necessarily held on a regular basis depending on the role of the group within the movements and its aims. Irrespective of the form that this type of politicization took, it was fundamental to the continued participation of a large number of women in the popular movements. Support groups or self-education groups were quite common in the popular movements and arose from a basic need for women to discuss issues linked to their political participation or to the political demands of the movement. The Unemployed Movement and the Health Movement provide excellent examples of how these groups functioned.

In the Unemployed Movement the formation of women's groups spread swiftly from one region to another and the numbers of women Committee members rose to 70 per cent. In some regions women's groups were formed as a result of initiatives by women themselves. In others, however, their formation was encouraged by men, who had organizational control of the Unemployed Movement, as a means of removing 'women's issues' out of the general political demands of the movement. Wherever they emerged, they became a forum for the discussion of a wide range of issues such as women's political participation, male domestic violence and women's role in society. Through the exchange of experiences, particularly those related to their own participation in the popular movements, many women were better able to face the problems arising from their political activities with the support of these groups which served to reinforce their commitment to the movement.[4]

Many women enjoyed the opportunity of talking about themselves and their lives, as well as finding out more about themselves and other women. These groups often developed into a mutual support collective wherein *companheiras* in the struggle became true friends who gave each other help in their personal, working and political lives. Some of these groups were little more than small women's meetings but others went on to join up with other women's groups and organize talks to which feminist speakers were invited to discuss topics such as 'The history of women's political participation in Brazil' and 'Female sexuality: my body, my choice'. Many of these groups joined with women's organizations and feminist groups on demonstrations and political protests about issues such as abortion, violence against women, and family planning. Consequently, these groups, although initially formed as a means of

politicizing women within the popular movements through consciousness raising undertaken by women for women, sometimes became a
vehicle for contacts with other women's groups or other women in a
wide range of political organizations. Irrespective of how these groups
developed, they gave women greater confidence in themselves and many
of the women who organized or participated in these groups went on
to be elected local and regional coordinators of Committees of the
Unemployed.

The Health Movement actually evolved from issues concerning
women's health and that of their children (see Machado, Chapter
4 in this volume). The use of self-education as a form of consciousness raising about these issues was a natural progression from the
movement's initial aims and objectives. Their struggle to improve an
inadequate and under-funded health service made women conscious of
the need to take action themselves in the area of preventive medicine in
an attempt to reduce the risk of health problems for themselves and
their children. Women of the movement were principally interested in
two main topics: first, ways in which they could prevent or reduce the
risk of their children suffering from some of the more common
childhood diseases in Brazil; second, access to information on contraception so that women could make an informed choice as to what methods
were available and most suitable for them to use. In both cases, women
turned to local doctors and nurses to help them produce booklets which
they could use in their self-education groups.

Many young doctors were more than willing to give technical advice
and information which the women could transform into easy-to-read
pamphlets with plenty of diagrams and drawings to facilitate understanding of the more technical information. In this way they were able
to produce informative and detailed literature relevant to their needs
and methods of self-education. Whilst the intitiative for the self-education groups came from the elected Health Councillors of the Health
Movement, the idea was enthusiastically approved by the women who
participated in the movement and swiftly attracted the interest of
women generally in the local community.

The 'course' on family health was based on the prevention of common
infectious childhood diseases such as dysentry, parasites and skin infec-
tions prevalent in poorer areas of the urban periphery. Inadequate and
often polluted water supplies, the absence of proper sewerage and
drainage systems and the questionable hygiene standards of local shops
selling fresh foods were some of the causes of the high incidence of
infectious diseases amongst children. Women discovered that it was
possible to take preventive action against such illnesses and in the
self-education groups they discussed ways in which they could help
themselves and their families on the basis of the information in the

pamphlets. Women learnt to sterilize their water supply, what to look for in fresh foods and how to identify certain diseases in their children to facilitate early diagnosis and treatment. The lack of emphasis within the Brazilian Health Service on preventive medicine meant that this form of self-education was extremely important.

The leaflets produced by the Health Movement on family planning were also an invaluable source of information for women. In Brazil there were no official family planning schemes or health advice services for women and the public health service offered little orientation for women who wished to use contraceptives, apart from advice on 'natural' methods of birth control supported by the Catholic Church. In practice, while there was no official line taken by the government on the question of family planning, women of the working poor were often pressured into using sterilization as a permanent solution to their 'problems' without being advised on alternative forms of contraception available to them.[5] The self-education groups and pamphlets were a means of informing women about the different methods of contraception available in Brazil, showing the advantages and disadvantages of each one to help women make an informed choice on what method was best for them. Women who did not participate in the Health Movement attended these informal groups on family planning and some of them went on to participate in the movement itself.

Meetings held by the Health Movement about the issues of family health and family planning were invariably held in people's houses, often as 'street meetings' to which women from the movement would invite their friends and neighbours. Working from the pamphlets, women would discuss the questions raised and any practical difficulties which arose. No professional people participated in the groups: they were run by ordinary women who sought practical solutions to health problems which were not resolved within the public health service. The existence of the groups demonstrated the importance of the Health Movement in the struggle for a better health service and they reinforced or developed women's commitment to the movement.

This type of consciousness raising developed for women by women raised female consciousness in relation to strategic gender interests. It stimulated and developed a complex matrix of inter- and intra-class alliances between women around gender specific issues. As women acquired a greater sense of themselves and gender inequality through their political practices in popular struggles, women's organizations in the low-income neigbourhoods emerged and grew in strength. By developing a political identity as women of the working poor they were able to define their relationships clearly with other organizations, both inside and outside institutional politics. Consequently, when they entered political protests in association with political parties and trade unions on the one hand, or feminist groups and women's organizations on the other, in general or gender specific political struggles, such as the Campaign for Direct Elections or Women against Violence Campaign, they were able to defend this identity and their practical gender interests from a position of greater political strength.

The demand for literacy and politicization

Adult literacy courses based on the methods of Paulo Freire, which use short literacy courses as a means of politicization, were also a popular and constructive form of conscientization.[6] Many of the urban poor who participate in the popular movements are illiterate. Women in particular have had few chances to educate themselves and when they begin to undertake organizational roles in the movement there is tremendous pressure on them to obtain basic literacy skills. Dealing with members of the government, participating in negotiations and the need to take notes at meetings present difficulties to those who are unable to read or write.

Women often bypass these difficulties by using tape recorders or by taking their children along to the regional or state meetings to take notes, so that they are able to recall the main points of discussion or proposals to report back to members of their group or the movement. However, the high rate of illiteracy amongst those who participate in the popular movements often gives a privileged position to people who can read or write. There are women with valuable political skills who have been elected as representatives or coordinators in the movements who hide the fact that they are illiterate. Whether it is from a feeling of inadequacy or as a result of internal or external pressure, it is invariably the leaders or representatives of the movements who instigate literacy courses as a means of educating both themselves and others in the popular movements.

Literacy courses are always undertaken with the help of activists from either the Church or the political parties who have experience in this field. These activists are invited by the movements initially to teach the literacy course but eventually one or two members of the movement are shown how to deliver the course and this eliminates the need for outside help. The courses are nearly always over-subscribed and it is frequently women who are the most interested in becoming literate. This interest derives, primarily, from the desire to enhance and expand their new-found political skills; to be able to make notes, to vote in government elections, to read political material or even to write their own placards were all skills that these women wanted to acquire. Thus, it was these factors which made this type of literacy course based on politicization ideal for the popular movements.

Community action as political action

Cooperative work and mutualist schemes were two other activities
which developed and strengthened the solidarity between those who
participated in the popular movements. Amongst the urban poor they
were a means of resolving immediate and socio-economic problems and
could be of a short- or long-term nature. Cooperative work schemes
usually entailed the sale of goods or commodities and only benefited
those who participated in them. Mutualist schemes, on the other hand,
involved the provision of services and frequently benefited those who
did not participate in them as well as those who did. Irrespective of the
schemes' beneficiaries they arose from the political practice of the
popular movements. These schemes became an integral part of their
political organization and extended the members' political commitment
to the idea of collectivism into the local community.

The basic philosophy was: what cannot be resolved until tomorrow
through political action will be resolved today by community action.
Although initiated by the popular movements, there were no barriers
to participation in them and many people of the community who were
not activists of the popular movements either participated in or supported these schemes. Hence, these schemes gave considerable support
to the activists of the popular movements in the 'low' periods of their
political organization by sustaining a sense of solidarity and commitment
to the struggle; they also attracted many new people to the popular
movements.

It was the Unemployed Movement that used cooperative work
schemes as an immediate solution to the subsistence problems of its
members and in the period 1983–5 they became popular amongst
the unemployed throughout Brazil. Some state governments financed
projects submitted by the unemployed but in São Paulo the majority
of financing came from the Paulista Association of Solidarity in
Unemployment (APSD) and was only given to groups registered with
the Association.[7] Once again it was women who were most interested in
the schemes which covered a wide range of petty commodity production
such as bread making, confectionary, tailoring and craft production.[8]
These schemes were 'tailor made' for women: they were located near
the home, they used flexible work rota systems and initially there were
few expectations that the financial remuneration would do more than
help to sustain the family unit during periods of unemployment. However, some of these cooperatives became extremely successful and the
share of the profits was comparable to women's wages in the formal
sector. Some cooperatives were more successful than others, in financial
terms, but most of them managed to provide subsistence wages for the
unemployed who participated in them. Furthermore, some of the

150

cooperatives gave a small percentage of their profits to the funds of the Unemployed Movement to help finance political action.

The initiatives for the cooperative work schemes were encouraged by the Unemployed Movement as a whole but individual projects were devised and organized by the local Committees of the Unemployed or Solidarity Groups of the Unemployed.[9] The cooperatives were organized along democratic lines and decisions were made by the collective about how it would be run. Those who wished to participate had to work a minimum number of hours per week in order to keep their place in the cooperative and receive a share of the profits. Anyone who did not comply with these regulations had to leave (there were usually plenty of people willing to take their place). In some cases people who did not participate in the movement were part of these cooperatives because the only qualification necessary to join was to be unemployed and many people eventually began to participate in the Unemployed Movement through cooperative work schemes. Relationships built up amongst the unemployed within the movement were strengthened by these schemes and the services that they provided benefited not only the unemployed but also the local community.

Mutualist schemes were, at one time or another, used by all the popular movements to resolve problems within the local community on a short-term basis. The popular movements found that considerable onus was placed on the urban poor themselves to find solutions to many of their demands such as creches, house construction, refuse collection and other services. The Favela Movements, for example, set up many mutualist schemes due to the difficulty of getting general services installed in the *favelas* and the time it took for *favelados* (*favela* dwellers) to gain these services through their struggles.

Between 1983 and 1985, the Favela Movement in Vila Sezamo organized a mutualist scheme for rubbish collection as well as for the construction of the movement's headquarters. In both cases, volunteers came from within and outside of the movement and even *favelados* who did not participate in the scheme gave money, tools or supplied food for those who undertook the physical work. The rubbish collection scheme was probably the best example of how these schemes tended to work.

Despite several attempts by the Favela Movement in Vila Sezamo to get their rubbish collected the local council had refused them an internal service. Rats, vermin and mosquitoes had created health problems and the movement proposed that the council send a truck to a point outside of the *favela* each week and the *favelados* themselves would deposit the rubbish. The council accepted their proposal and promised to provide placards to be posted around the *favela* to prevent dumping and to install large litter bins at various locations within the *favela*.[10] The mutualist scheme arose from this proposal. Part of the scheme was to

151

educate people to take a pride in the *favela* and keep their environment clean. Members of the Favela Commission held consciousness-raising meetings about the importance of their scheme to people's health and its benefits to the community itself. They also made house-to-house visits telling people about the scheme and asked for volunteers or practical help to make it a success. The scheme was the first of its kind and its success encouraged the Favela Commission to introduce other kinds of schemes, including cooperative work projects, to benefit the *favela* community.

ORAL HISTORY: THE STORY TOLD AND RETOLD

The use of oral history as a form of politicization was fundamental to the creation of a political identity for the movements themselves and for those who participated in them. Many of the popular movements do write down the history of their organization and struggles and use this material for the conscientization of new members. But the use of oral history is a rich and personalized tribute to past events and the contribution of each of the participants to the movements' 'success'. An oral history is developed from individual and collective experiences of those who participate in the movements. It is a vivid, living testimony of their political defeats and victories which are recalled at any and every opportunity to demonstrate the courage, tenacity and commitment of those who participate in them. Women, without a doubt, are the most avid subscribers to this form of politicization. They take great pride in their ability to recall events and even conversations in the minutest detail, often dramatizing the conflicts and confrontations with government or the police during their struggles. It is the way in which they reaffirm the importance of their participation in this form of political organization and the viability of the popular movements as an instrument for political change and social change.

In the meetings held after a political protest has been carried out, those who participated recall their personal experiences of the struggle and how they felt about it. Not everyone who participates in the popular movements can or will participate in their protests and demonstrations. The practice of using individual accounts of events informs those who were not present but it is also a means of encouraging far more people to participate. Everyone is given a chance to speak or ask questions; the emphasis is on the individual contribution to the collective; the relaxed, enthusiastic atmosphere of these meetings alleviates the inevitable anticlimax which follows the intense period of political activity leading up to collective action. For those who participate in the popular movement *every* struggle is a success which is counted not only by the concrete victories obtained through the struggle in relation to the demands of the movement but also by its impact on both the government *and* the participants.

These meetings are an emotive and exciting vindication of the latest political protest organized and carried out by the popular movements. No one seems to tire of the inevitable repetition within individual recollections; each one is different because it discloses a personal interpretation and experience of the protest. Each moment of the protest is relived and retold in an atmosphere of excited interest and anticipation which conjures up an intricate web of mental pictures in the minds of those who listen attentively to these recollections of the latest act of political defiance. Individual acts of courage are recorded and acknowledged in these meetings; even those who suffered personal hardships through their participation in the protest are pointed out and applauded. It is here that the history of a movement is recorded and developed not through the eyes of an observer but through the recollections of political actors themselves.

The oral history of the popular movements catalogues both the triumphs and tribulations of their political practice, and those who participate in the movements are the ones who make and develop this history. The past successes of a movement are often what sustains its organization and the political commitment of those who participate in the movements in some of the more difficult and less successful periods of its political action. Women in particular are eager to record their political experiences, to create for themselves and reaffirm a specific political identity, one in which women are not inconsequential but successful political actors. In this way, oral history is not merely an adjunct to the political action of the popular movements but one of the key elements of their political practice.

THE TRANSFORMATION OF GENDER INTERESTS

The many forms of political action employed by the popular urban social movements, as we can see here, have many different functions. The need for conscientization or the creation of solidarity within the movements themselves was not always the initial purpose of these practices but often became the reason for continuing or developing them. Some practices arose from a simple need or idea expressed by those who participate in the movements whilst others were a direct attempt to conscientize the popular classes. Irrespective of the impetus for or the development of such forms of political practice used by the popular movements, in most cases they reflected the desires and needs of the women who dominated this form of political organization. In doing so, it gave many women the chance to 'improve' themselves and develop both socially and politically by utilizing a wide range of skills to enhance and expand their new-found political ones.

The political mobilization of low-income women arose from their practical gender interests as well as structural class differences in Brazilian society. Their daily battles for economic survival prioritized political action around issues related to the access of the popular classes to the benefits of economic development. The provision of urban infrastructure, adequate healthcare and transport not only affect women's activities in the reproductive sphere but also have limiting effects on their access to employment and income generating activities. Nevertheless, as opposed to gender 'neutral' analyses (Slater 1985), it would be wrong to describe urban social movements as women's social movements (Safa 1990), this would fail to acknowledge not only the participation of men in these movements but also the affects of gender relations within and between these movements and institutional politics.

In the popular movements men can account for up to 40 per cent of the participants. Gender relations in this context are different to those in institutional politics. Women acquire political experience in association with men, but in a sphere where they predominate it is frequently on their own terms. As we have seen here, the political practice of the movements are strongly influenced by women and as such their political development is directly related to their gender subordination in society. They are able to strengthen and legitimate their political role and become experienced political actors. Those who go on to enter institutional politics are able to do so from a much stronger position and with greater confidence in their political abilities (Corcoran-Nantes 1988; Moser 1987).

Through their struggles around practical gender interests women who have a similar socio-economic experience in Brazilian society develop greater solidarity and awareness in relation to strategic gender interests. Opposition to women's political participation at a personal political level reinforces their experience of gender inequality in other spheres. Through their contact with political parties and trade unions they develop cross-class links with other women from feminist groups and women's organizations. These links have been strengthened by their association in struggles around strategic gender interests. By developing a political practice which emphasizes not only class inequality but also gender inequality, these women began to construct a gender identity around strategic interests based on their socio-economic experience. Consequently, low-income women were able to articulate their priorities and interests in relation to other class-based feminist groups and organizations and to pursue strategic gender interests through their political action.

Nevertheless, for low-income women practical gender interests take priority in their political struggles and it is here that they have built the necessary basis for unity and solidarity. Class oppression, to which their gender subordination is directly related, has forced women to organize around issues related to their very survival and that of their families. These issues comprise the major social and economic problems in developing countries and in this context strategic gender issues take a secondary role or may not be considered at all. In Brazil, however, amongst women of the urban poor female consciousness has developed around strategic gender interests, and whether they choose to describe these as feminist or not is irrelevant. What is important for women of the popular classes is that their concerns are firmly on the political agenda.

7

TOUCHING THE AIR
The cultural force of women in Chile

Catherine M. Boyle

Solidaridad, es bonita esa palabra, es como nombre de mujer.
(Solidarity, it's pretty, that word. Like a woman's name.)
(Taller de Investigación Teatral 1978)

LA CUESTION FEMENINA

Women's right to vote has always seemed to me to be a most natural
thing. But, I distinguish between right and wisdom; and between
natural and 'sensible'. There are rights that I do not care to
exercise, because they would leave me as poor as before. I do not
believe in a women's Parliament because I do not believe in men's
Parliament.
(Mistral 1928)[1]

In 1927, when Gabriela Mistral wrote these words, women in Chile were
fighting for the right to vote, which they did not win until 1949.[2] On
the surface, these words would seem to be anti-feminist, but, in fact, they
point to something that is fundamental to the feminist debate in Chile,
and that is why I have chosen to start with this quotation. It is indicative
of what, from the outside, would seem to be an insurmountable
contradiction in attitudes to women's rights and role, yet it demonstrates
a specific culturally based and constructed view of womanhood.
Throughout her work and life. Gabriela Mistral sought a meaning to
the idea of womanhood, of femininity, of maternity, and she sought,
likewise, to understand all the ways in which she herself was not accepted
as woman, being neither married nor mother. What Gabriela Mistral
was trying to do in 1927 is reproduced in the aims of later generations of
women in Chile. She was searching for ways of being female that were
not pure negation, that were not a continual act of asserting in the face
of negative values, in the face of being defined as everything that was
not male: ways, that is, that expressed the positive, the active, the
possibilities of women, whether won and fulfilled, or lost and abandoned. What, in the passage quoted above, she seemed to deny women,

was, in effect, the falseness and weakness she deplored in the rule of men.

She looked, as she said in her famous poem 'Todas íbamos a ser reinas' (We were all to have kingdoms) (Mistral 1938) to name the 'kingdom' of each and every woman. In this poem she recreates the little girl's dream of being princess over a kingdom that would lead all the way to the sea, she talks of the ways in which that dream collapses, and of the eyes that turn black 'for never having seen the sea', and of the little girl who only found her kingdom 'in the moons of madness'. Yet the dreams live eternally on, ever renewed by new generations, renewed and strengthened, and leading surely to the sea, to the open, to freedom.

Gabriela Mistral named something of fundamental importance for women. She named the right to self, to dominion. That is crucial, and is at the heart of the impetus for expression, cultural and political, of women in Chile today, who are now naming themselves, and whose 'kingdoms', like those of Gabriela Mistral's princess, are born from the imagination, go beyond the self and the confines of their physical limits into politics, into what has too long been referred to as public life. Women in feminist movements look to transgress the boundaries of power, but by being women, by acting as women, by demanding as women and as citizens. They are looking to neutralize the idea of public and private realms. They are moving the political parameters, placing themselves in the centre, in multiple centres of political life; they are renaming the problem, eliminating the notion of the 'woman question', which has long been defined as a problem, and posing the real problem of the gender question, which goes deep to the roots of society.

In this chapter I want to look at the way in which this shift is founded in both the cultural and the political, how it is given life in the cultural expression of a political experience. I want to do this by looking at political manipulation of images of women through which women, as mystery, as unknown, are problematic, but as mother, as solid, eternal, unchanging, and culturally definable, are on one hand wooed and on the other feared. I want to study how the control of these images is, in effect, a usurping and a subversion of roles through which women in reality identify themselves, but in ways which are forceful, active, public and positive. First, I want to look at aspects of the development of this process.

In Salvador Allende's Chile (1970–3)[3] the 'woman question' was acknowledged as being important, women were to be part of the revolutionary process, harnessed to the cause, working alongside their menfolk, and guided by them on Popular Unity's 'Chilean way', the peaceful road to socialism. This placing of women in the process of political change is perfectly represented in the song 'Venceremos' (We shall overcome) by the group Quilapayún, which became a form of anthem of the left, and where women were included in the fourteenth verse, which says, 'y la mujer de la patria también' (and the woman of the fatherland as well).[4] This allotted place of women, under the wing of the real agents of political action, is amply illustrated in Isabel Allende's *The House of the Spirits* (1982) whose Popular Unity heroine is guided in political thought and action by her lover, and who devotes herself to the political struggle only as a result of love. This protagonist is the perfect literary representation of the ideal Popular Unity woman. She is a woman first, political agent next. She is unobtrusive and politically well behaved.

Salvador Allende knew that women were important in Chile in terms of consolidating his shaky electoral success. Since women won the vote in 1949 voting patterns had shown an inclination towards conservatism, although, as we shall see, the analysis of women's votes must be put into perspective by comparing it with the way in which men vote (see Zabaleta 1986). Allende believed the female vote to be significant, and he set about wooing it. In speeches tuned for their ears, he adopted the tones of the supposed female language of dedication and sacrifice in order to reassure women of his identification with their role, and he underlined that there would be no violence in the process of change, only building and creating (see Chaney 1974: 269). He wooed Woman in this role, trying to win over her loyalty, tempting the potential mother – in the guises of wife, lover, companion – to share the revolutionary goals of Popular Unity, which, on their realization, would carry within them the goals of women, 'in essence born to be mother' (ibid.). He encouraged the men of the coalition to bring their womenfolk out of the home, into the streets, to demonstrations and marches, to show them the riches of public adherence to the political struggle and the revolutionary process. And women did participate actively, as represented in Isabel Allende's protagonist in *The House of the Spirits*, or in the play *Nos Tomamos la Universidad (We've Occupied the University)*, about the student role in the University Reform of the late 1960s.[5] Yet, they are always alongside, rarely directing, or at the centre. Popular Unity did, in fact, set out programmes to deal with childcare, nutrition, maternity provision, all aimed at incorporating women's needs and demands into the process of revolutionary change. But, like so many other aspects of the programme, these were beyond the possibilities of the government at that time, given the constraints of working within the existing constitution, the difficulties Popular Unity suffered from the very beginning, and, perhaps more importantly, the huge cultural changes that would have had to have taken place before any of these aims were realized.

Women were important for Allende and Popular Unity, but they were perceived as important from an externally and eternally imposed definition of women. Women were a mass of beings, externally known and identified, with needs that, in theory, politicians should easily identify and satisfy. But beyond this identification as a problem they were all but ignored, and subsumed into other groups. Perhaps, as Chaney suggests, there was no real 'conviction that gaining women's support outside the electoral sphere' was important, or perhaps women were perceived as a potentially destructive force:

> By mid-1973 Chile faced many urgent problems: food shortages, high inflation, miners' strikes and falling copper prices, lack of foreign exchange. Perhaps there simply was no attention to spare for solving the woman problem.
>
> Or perhaps there was an underlying fear to see women mobilized at all. The famous march of the empty pots organized by middle class women to protest food shortage in 1971 had reverberated around the world. All through the waning months of 1972 women were beating their pots again in the *barrio alto* (the middle and upper middle class districts of Providencia and Las Condes). Women of the Right, encouraged by their success, were busy organizing a movement across party lines, Poder Femenino (Feminine Power), to shore up the opposition forces and to 'fight for our homes and families'. Perhaps there was an unspoken distrust that women, even of the left, could be controlled, and a disinclination to facilitate their mobilization.
>
> (Chaney 1974: 271)

Allende knew, from his experience with other sectors, that mobilization carried its own, potentially uncontrollable, dynamic. Although, in general, he could not or would not do anything to control them, he was aware that there were certain sectors whose rapid political incorporation was a serious threat to the ideal of containing the revolution within the legal parameters of the constitution. Yet women were still left as a subgroup of these sectors. Until the empty pot demonstrations showed how gender specific politics could be effectively manipulated, women were not seriously regarded as a sector within society with specific needs or demands, or an active political potential.

So, Salvador Allende was conscious of the importance of the female vote, but he never won it (ibid.: 268). Why? The simplest answer is that he was regarded by the women of the middle and upper sectors as a threat to the order of society, to long-established moral codes, and to the sanctity of the family. The simplest answer is fear. But the sense of threat and fear during these years was not confined to women. The female figure became the dramatic representation of this fear in Egon

159

Wolff's (1971) play, *Flores de Papel* (*Paper Flowers*), whose protagonist, Eva, is enticed into a new aesthetic, dragged into emotional and financial ruin by a noxious tramp, and paralysed into impotent compliance with her degradation. Fear of the threat of the success of the Popular Unity model – seen in terms of destruction and disintegration – was a vital part of the middle sectors' support of the military *coup* of 1973.

Without further investigation, it would seem ironic that it has been during the years of the Pinochet regime that women have most effectively created the space for independent voices, and it has been in the transition to democracy, following Pinochet's fall in 1988 and democratic elections in 1989, that these voices have been most forcefully heard. In 1988 Pinochet finally followed his own mandate in the constitution of 1980,[6] and held a plebiscite in which he was the only candidate, and in which there were two ways of voting: Yes (in favour of Pinochet) or No (against him) (see Angell 1990). The outcome was absolutely crucial for Chile. If Pinochet won the plebiscite, the left and centre would find themselves totally disarticulated, and Chile would lose sight of a return to democracy until at least 1997. In this plebiscite, the voices of women were heard loudly, both for and against the only candidate and, on both sides, firmly linked to the official campaigns for the Yes and for the No.

Pinochet and the women of the right made constant reference to the, by this time almost mythic, role of women in the downfall of Allende. When he addressed the Movimiento de Mujeres por Chile (The Women's Movement for Chile, a grouping of women around Pinochet's wife, Lucía Hiriart de Pinochet), the General – and soon to be only candidate – recalled those who 'deceived the nation with their "Forty Measures"',[7] he declared his reliance on the support of women, so that the same people would not return to repeat the same disaster, 'to destroy the nation'. In the same speech he noted how women had registered to vote in greater numbers than men, and how more than 30 per cent of these women were 'dueñas de casa' or 'mujeres del hogar' (housewives, ladies of the house). He continued his address in the following terms:

> I am confident that this meeting shows strength, the strength that you submit to the person who at this moment is responsible. This strength that makes our belief in Chile live again, the belief that we must triumph.
>
> In Punta Arenas this morning I said to the women: you who have a sixth sense, which is to smell out danger, have to have faith in the fact that you can destroy walls in order to triumph. This faith I see in your hearts. That is the faith that gives me the strength to carry on.

He addressed woman as mother, as keeper of the faith, as possessor of a purely feminine sixth sense, in terms of the loyal wife. It was only in passing that he acknowledged the base importance of the electoral power of women. Pinochet cast himself as father and husband, he sought the submission of female strength to a cause for which he held absolute responsibility, and he called for the faithful support and unrelinquishing devotion of each individual woman.

The Mujeres por el Sí (Women Vote Yes), confident in their powers for the moral guidance of the nation, published an announcement in the press, urging the correct vote and warning against a return to chaos, anarchy and, they said, bloodshed:

> Chilean Woman: We address you, again, as one of the first to express bravely her adherence to the path that Chile has followed in the last fifteen years.
>
> This time we will be even more bold and declare what we denounce. **Chile is today facing two alternatives**, one has always been immutable, permanent, that is the one our present Government presents, the other has disguised itself on a number of occasions, at first it used terrorism in a fierce and steady manner, when it saw how firmly our people rejected it, it eased its tactics, it began to look for companions dressing itself up in sheep's clothing, and many have fallen into this trap, and today, by deceiving some, – and through some who allow themselves to be deceived? – they talk of democracy, trying to make bread without flour or water, hidden behind the comfort of those who ignore the following points.

> (Mujeres por el Sí 1988)

The declaration continues in the moral linguistic codes related to traditional perceptions of Christianity and democracy. It warns against the degrading and denegrating influence of the 'wolves in sheep's clothing' mentioned in the paragraph quoted, and declares the intention of persevering in the march 'under the present government' towards peace, tranquility, progress, and a well-deserved place among the leading capitalist nations of the first world. This language firmly reproduces the limited Pinochet repertoire of images and ideological assertions.

On the same day as this declaration, the staunchly pro-government daily newspaper *El Mercurio* published the findings of an opinion poll which they declared had proven again the right-wing leanings of the female population.[8] The very fact of the poll points to the identification of the female vote as an issue. On 22 August 1988 the left-wing paper *La Epoca* had said in an editorial: 'Women will have the last say in the forthcoming plebscite'. But here, the author attempts to show the complexity of the question, probing the reasons why women have 'received historically the role of primordial defensors of the private realm', and what this has actually meant in electoral results. And, again, the voting patterns are shown to be complex, tending more towards attitudes that reject 'the instability inherent in radicalization' than to an overall vote for the right. Both of these pieces reveal, and, indeed, advise the targetting of women as a sector to which attention and care must be paid. The 'woman question' was most certainly in the air.

It was in the air because of the perceived role of female mobilization in the downfall of Allende, which was part of the collective memory of the last experience of the democratic process. What this memory represented was the potential stridency of a cross-section of Chilean women, among whom were those who now looked to the 'immutable', to the 'steadiness' of their own role and to the figure of Pinochet to keep them 'on course'. But their voice was not the only female voice, for during the Pinochet years there had emerged the articulation of independent women's voices from the left, linked to the political parties of the left, and instrumental in creating a separate women's agenda for change.

Concertación, the coalition of parties for the No vote, also addressed Chilean women, in advance of the plebiscite and anticipating the role women could play. These are the first two paragraphs of their political agenda, published in *La Epoca*:

Today, when the whole country is preparing to take decisions that are fundamental for the future, women want to have a say. We, feminist women, conscious of the moment that we are living, want to propose that all women, young and mature, organized and independent unite as women, in order to express our own demands for Democracy and demand that these form part of the democratic project to which the majority of us aspire.

In what follows we propose a set of demands that we believe are strictly just for all women. We invite women to support them and to present them together to the democratic parties, to social and trade union organisations, to religious institutions and to the Chilean Commission for Human Rights, so that they receive their commitment that our demands will be an integral part of the contents of the democratic system that all of us, Chilean women and Chilean men, will build.[9]

The authors of this statement were women who were committed to Concertación, whose public face was, however, almost entirely male. Though actively involved in Concertación, they presented their demands of the new democracy as a further agenda. For the first time since winning the vote in 1949 (see Kirkwood 1990: 131–75), there emerged a voice for specific female demands, rooted first and foremost in the role of woman as citizen, not primarily as mother and guardian angel of the household, and setting men alongside them in the fight for equality, for real democracy.

Here women made two explicitly challenging statements: first, that the 'woman question' was not an issue apart, it was a gender question, a question of readjustment of attitudes and roles; second, they declared that the return to democratic rule did not, in any way, guarantee that women would live a life of public and domestic democracy, free from varying levels of male authority, violence and abuse. No government had ever guaranteed that, no government had ever regarded it as their responsibility to identify and respond to the specific needs and demands of women within the home. Julieta Kirkwood had pointed this out in the following terms in 1982:

> If, before, women did not wholly value the meaning of liberation and had accepted an inferior type of integration, now, in the face of authoritarianism, they are, to a certain extent, confronting a familiar phenomenon: authoritarianism as a culture is their daily experience. The return to democracy will not be for women the reapplication of the recognized liberating model.
>
> (ibid.: 186–7)

This is profoundly important. It articulates a previously unnamed perception of the realities lived by women. These realities are of a hierarchy of subordination to a series of ruling male values, of living within patterns of behaviour that are externally imposed, based on unquestioning notions of female identity and role. They are realities that create boundaries within which the individual adopts forms of behaviour and expression that will not seriously compromise the fragile edifice of the structure that sustains domestic survival. As Jorge Gissi Bustos (1980: 36) says, this is based on the 'dominant ideology . . . that nature has given authority to them [men] and they act accordingly'. He notes a system of interchange in this tacit deal, and he claims that it brings in its wake 'the apprehension of losing the advantage of feminine submission given in exchange for masculine protection' (ibid.: 37). There is a clear relationship between the sustaining of this agreement, and the fear of losing its deemed advantages, a fear that can result in a deeper entrenching of authoritarian patterns of control within the household. It is the awareness of an enduring and unacknowledged authoritarianism within the home and family structure that informed the demands of the feminists of Chile in the agenda quoted here.

Naming the problem in this way re-routes it, away from the women question, in the direction of the gender question, towards ways in which both sexes are confined in these patterns of domestic and social authoritarianism. The question, they state, is one of liberating untapped forces in society, and so a question of monumental changes on

163

very level of the organization and distribution of labour, subsequently accommodating the consequences of these changes.

The demands set out by the feminists of Chile in 1988 as part of a possible agenda for Concertación hit at the core of social, political and cultural definitions of women. They indicate the need to look at the reasons for certain definitions of women, and especially the establishing of one role for women, that is, as first and foremost mother. I believe we find two basic consequences of this attitude, which, of course, have many wider repercussions. First, it gives legitimacy to women's being, and second, it establishes an identity like a monolith, solid and, as the Pinochet women boast, immutable. Yet, it is that very stating of the role of women as first and foremost mother that makes the role political, especially in historical circumstances like those in Chile, where, in one period the family is perceived as being under maliciously calculated, heathen attack, and in the following period, the family, held aloft as a sacred unit of society, is in many sectors in a period of violence, repression and disintegration. I want to suggest here why the role as mother is fundamentally political by returning to the notions of fear and threat.

The Pinochet *coup* came about partly as a result of the perceived threat and fear of looming chaos, economic collapse, imminent social unrest, destabilization.[10] For the right wing, the country was ailing, nobody was free from the treacheries of the Marxist regime. The political became private in important ways, for with food shortages and their counterpart – panic storing of the basics by the stricken well-off – the home was invaded by social change, by the public threat. And it was then, when, they said, their cupboards were bare, that the enraged women of the right took to the streets with their empty pots, and often with their maids. The home had been, both metaphorically and really, invaded and violated, and they declared openly that it was the family that was most at risk. The family was identified as the place where the weakest in the society reside and remain, expecting and needing the protection of those, or, equally, against those in positions of public power. The wives of the powerful transport workers, who crippled the country with strikes in 1973, used this image explicitly in their pamphlets:

> We don't have bread for our children! . . . 'We don't have a roof to put over our heads! . . . We have been violated, humiliated, and hounded because we have defended our sons, because we have shown solidarity with our striking husbands, because we have gone into the streets to arouse the sleeping consciences of so many men!'
> (in Mattelart 1976: 285)

The family was, in these terms, women and children, the mother the moral centre both of the family and of the 'sleeping conscience' of the

164

nation. In the sinking ship that Chile in 1973 has so often been depicted as, women were to be its saviours.

The manipulation and evocation of the family for calculated political gain is the conscious manipulation of the social truth that the family is at the core of public stability, that the certainty of the easy continuity of the external rests on the security of the domestic unit. And its very core, in this scheme, is woman, the mother, the final, the strong, the central axis of society. The representation of women as the supporting pillar in a socially convenient construct is absolutely fundamental to the maintaining of a series of levels of certainty, no matter what the regime may be. And this is why women are at the core of a sense of threat. They may fear for their family, honestly and legitimately, but they are also in a uniquely powerful position when the home is consciously set up as the last bastion of a secure social order.

It is because of this representation of women that the private – home, children, responsibility for the household – is political. And it is through cultural expression that we give and are given the ways of seeing and understanding how the political experience is lived on the most private of levels. In the following section, I want to look for ways of exploring the interaction of the political and the private in terms of its concrete expression in the cultural. But in order to discuss this, I want first to suggest a simple working definition of culture.

THE CREATION OF THE CULTURAL

Culture is how people express their daily lives and how people express themselves in their daily lives. Cultural expression is the *material* expression of the interaction between an internal self and experience with an external reality. Culture cannot be seen as an immutable solid apart from or pinned on to society, for it is integral to society, it grows from it and feeds it, it elucidates its day to day workings, and provides ways of talking about ourselves within our different contexts. Without cultural expression a society is void of forms and means of reference. It is void of concrete evidence of the meeting points between the inner self and external realities.[11]

When the upper-class women of Santiago's *barrio alto* took to the streets with their empty pots their motivation was anger: the welfare of their domain had been seriously compromised, the comfort of their domestic set-up destroyed. The government had failed them on the most basic of levels, the level of nutrition. In this instance, the pot was not a kitchen utensil, it was a symbol, a representation of basic human provision; the empty pot was the representation of the failure of the state to satisfy a basic need. This failure on the part of the government resulted in the inability of the mother to carry out a key role, that of

165

caring for her family. The empty pot did not only mean physical need, it meant a role threatened, it meant crisis in the household. The demonstrations, orchestrated, as Mattelart (1976) shows, with almost military precision by Poder Femenino, meant that this perception of reality and its penetration of the area of political action heralded definitive trouble for Allende: women on the march augured badly. As Julieta Kirkwood pointed out, the demonstrations prompted renewed worried questioning in the left about women's so-called conservatism, about the direction their political participation had taken (Kirkwood 1990: 49–50). These demonstrations were a shock. Women who had been regarded as apolitical stepped beyond the bounds of decorum of the middle-class housewife and occupied the streets. The kitchen, the home became political. This invasion of the public by the private had a lasting impact on the collective memory and imagination, providing images that, as we have seen, were evoked in 1988 in the public declarations of the Mujeres por el Sí, in Pinochet's speeches, in newspaper articles. The empty pot had become embedded in the collective memory as a political symbol.

This was clearly seen in the movement for democracy, and the plebiscite campaigns. In the early 1980s, the space for political expression began to open, people began, they said, to lose fear and to take up political action to rid themselves of Pinochet, who, naturally, fought back. In 1985 he imposed a state of siege, and Chile experienced one of the worst periods of repression since the mid-1970s. Protest was, once again, sent underground. Yet, throughout this period, people subverted Pinochet's imposition of silence and civic invisibility in order to make a voice heard. This voice found representation in the noise of the pots.

At a given hour on days of protest, people would go outside, into backyards or gardens, and bang pots. In the *poblaciones*[12] the din would be unbearable, in some areas of the centre of Santiago, the same would be the case, but the noise died out to the timid sound of the occasional solitary pot towards the middle- and upper-class districts. A past symbol of right-wing opposition, a symbol of the downfall of Allende, was consciously and successfully subverted. And behind the use of this symbol lay the same reasons, albeit of a very different magnitude. The long-term economic policies of the regime, which had benefited large numbers of the middle and upper sectors, had caused huge unemployment that had hit the *poblaciones* hardest. Empty pots did not mean, in this case, not being able to serve the dish of your choice. Empty pots meant no food, eating tea and bread, buying *calugas de aceite* (caramel-sized quantities of cooking oil). In simple terms, an empty pot meant not being able to *parar la olla* (fill the family pot). Pinochet, who had trusted that women and their domestic trappings – such as pots – would, after the *coup*, return indoors to a designated docility, had, in these

166

sectors, achieved exactly the opposite. Repression, disappearances, high levels of malnutrition, infant mortality, family breakdown, growing prostitution including unprecedented levels of teenage prostitution, had all meant that the private sphere of the domestic was constantly invaded and undermined by the consequences of socio-economic policies. Women for whom the family was the core of their lives, for whom the family was the central and uniting element in their community fought to defend it. And again, out of a reality of daily existence, came its physical expression in an expression of protest. The pot had been used as a vehicle of protest by the right, but the very fact that the pot is part of every single household means that it is a potential symbol for all. Like all symbols, it belongs exclusively to nobody, yet to everyone, it can be and is consciously used to store up memories and, like all symbols, it is open to subversion or appropriation by other groups. In the 1970s a network of signs was created that was linked primarily with testimonial forms, and women were at the centre of this process.

Women had, since the mid-1970s, depicted in testimonial tapestries, *arpilleras*,[13] the *olla común*, the common pot, a type of soup kitchen set up by neighbourhood organizations, often in collaboration with Church solidarity groups. Along with the solitary woman with a picture and a name pinned to her breast, the disappeared person, *población* houses, the Andes mountains, the *olla commún* became an integral part of the *arpilleras*, as a standard depiction of diminished resources and the diminished ability of parents to look after and protect their children. At the same time, in song, in poetry, in theatre there were constant references to a private sphere that was constantly intruded upon by raids, by the sound of circling helicopters, by gunfire and sirens, by the 'war' being waged by Pinochet that constantly came into their homes.

It is in their recounting that these experiences become part of a cultural expression. The *arpilleras* are a store for the memory; the women who make them, the *arpilleristas*, are the guardians of the memory of the *pobladores* in the years when any overt, verbal expression of this memory was disallowed them, and when any real and visible social existence was denied in the interests of the 'tranquility' of society. This process of remembering and storing is simply illustrated in the following extract from the play *Tres Marías y una Rosa*, where the newcomer to a workshop where *arpilleras* are made is sure that she has no experience to speak of, and nothing she can express in a tapestry:

MARUJA: And do you know how to make *arpilleras*?
ROSITA: I've only seen Señora Luchita do it. They're lovely the ones she makes, aren't they? So colourful.
MARUJA: Yes, this work is lovely, Rosita.

ROSITA: Of course.

MARUJA: Look, the first thing you have to do is think of an idea for a theme.

ROSITA: A theme.

MARUJA: Things that have happened in your life, that's what the theme should be. Because people are interested in what happens in your life, Rosita. We think that they aren't, but they are.

ROSITA: I don't think I'll be able to.

MARUJA: You've got to look back and remember something that's happened in your life, part of your own history, do you understand?

ROSITA: No.[14]

Rosita is lost until she can recall a time when her family were forced to eat only cabbage and salt, a recollection that is turned by the older *arpillerista*, Maruja, into a 'theme', the theme of poverty, of hunger, of the impact of the recession. For this character, as for many, the step beyond the private is taken almost imperceptibly, there is, initially, no clear appreciation of the fact that personal experiences have a relation to wider, social problems. Yet, once it is related to that wider social reality, and given a name within that context, it does not leave the private realm, for it is still rooted in and experienced from that realm. In the *arpilleras* experiences that remain part of an individual reality are expressed through a domestic medium, sewing, and elaborated for public consumption among a group of women who share, communicate, remember, create, and look to the building of a future. The private is the pool, the source, and because of the nature of the themes, laid in the laps of these women by political events, the private is political. There is no contradiction, the link is natural.

The *arpilleras*, at first, were naive depictions of a desperate reality. They were statements without words, a cry against the absolute invisibility of the hell of the *poblaciones* in the worst years of repression after the *coup*. At first, artistic merit was not an issue, figures were simple and told an obvious story through pictorial representation where words or explanations did not figure prominently. As time went by, however, the meaning of the *arpilleras* changed. The women who, like the character Rosita, had no sense of themselves at the outset as commentators on their society, or as artists, began to see themselves and talk about themselves in both these roles. The *arpillera* – largely invisible within the country, still, in the last years of the dictatorship, hidden underneath counters and only produced for sale on request – became, more than a testimony, the deposit for memory, for ways of expressing and talking about those years, ways of saying without using words.

The women of the *poblaciones* became a reference point for other women. This was partly because of the severity of their experience, which was seen as an extreme of experiences lived by other women. But there were other reasons that go beyond the question of gender identification and solidarity, to a wider question of how different sectors or groups talk about themselves in relation to political events and social realities. For example, in theatre, the most dynamic, communicative form of cultural expression of the Santiago middle classes, marginality was one of the principal themes of the seventies. On one level, this was part of a journalistic, testimonial role taken on by those in the theatre, but it was more than that. Talking about marginality was a way of demonstrating unease about the new order in Chilean society, it was a way of naming a focus of anxiety among those who lived in, sometimes benefited economically from, but did not accept the Pinochet ethos. The same dynamics are at work within the women's movement. The experience of the *pobladoras* is vitally important not only because it is one of the most extreme examples of social, economic and cultural deprivation, but also because it is a reference point. It is a form of mirror in which are reflected, magnified and distorted experiences that many women would say they share to some degree.

Memory is at the core of this shared experience. When women create a contemporary memory, they also create an awareness of a lack of a memory that projects into a collective past. As Julieta Kirkwood points out, women in Chile did not, in 1973, possess a specific political memory (see Molina 1990: 85). After the winning of the vote, women all but disappeared as they were integrated into the existing political parties. Women joined the long tradition of the 'well-behaved feminist' (Kirkwood 1990: 91). And men remained in possession of the social space and so of the enunciation and telling of the past.

For the Pinochet regime a political memory, specifically of the left and of the Popular Unity years, was dangerous, and was to be eradicated in order to 'cleanse' society. Men, in the unchanged perception of gender, became the principal victims of a concentrated process of physical elimination of this memory and past. In this way men became a crucial factor in the articulation of the link between the domestic and the public, the private and the political. At a moment of male disarticulation, the people who were in a position to rebuild, to rearticulate, to give a voice to experience and dissent, were women, especially in the *poblaciones*, where raids could clear a *población* of the entire male population between the ages of 15 and 55, even if only to hold them in a sports ground for a few hours.

As the multiplicity of women's movements grew in the Pinochet years,[15] there was at their centre a great feeling of creation, of building, of establishing bonds with a past that was only now being unearthed, and giving women a sense of past, a sense of history, a sense of memory.

LO QUE ESTA EN EL AIRE: WHAT IS IN THE AIR

In 1989, at the time of the plebiscite, there was a demonstration by 2,000 women in Santiago.[16] The women walked in silence along the main avenues of the centre of Santiago, carrying cardboard cut-outs of disappeared people, each one with a name. No words were spoken, only the image was seen, uncontaminated by slogans, for by that time the word was not to be trusted in Chile. The function of the march was to save the disappeared from a further disappearance: elimination from the collective imagination at a time of public euphoria, the time 'when happiness was coming', as the campaign for No proclaimed, or 'Spring was on its way', in the words of the campaign for the Yes, which usurped one of the most popular symbols of the return to democracy. The women knew that it was fundamental that, with the return to democracy, human rights issues, the fights of the groups for the detained and disappeared, should not be pulled into the euphoria of the success of the No and forgotten or abandoned as a threat to stability. The women knew that elimination from the collective imagination was the first step to elimination from the collective memory. They were no longer the 'well-behaved feminists' of before, they were aware of the continued workings of authoritarianism. And if words like 'democracy', 'stability', 'tranquility' and 'peace' were now in the mouths of the centre and the left, that did not mean that their meaning was strong and honest. It meant, rather, that these words had to be reinvested with pure meanings, meanings that would remember the past and make way for the future.

The women who organized these demonstrations were poets and writers. They were in tune with what was in the air in terms of the collective preoccupations in society in the process towards democracy, whether explicitly stated, or relatively unarticulated. They knew the key symbols of cultural expression under authoritarianism, and more than that, they had been part of the building of a system of elaborating and diffusing these symbols, of speaking and saying, keeping alive the languages of memory and the telling of experience. Women had transcended the private level to be an integral part of the naming and expressing of the present, and of a dynamic impetus for the future.

Sometimes it is difficult to give a specific, documented reason for what is happening, for the consciousness of it is little more than an awareness of something that is in the air. In 1988, as I shared the experience of the Chilean people returning to democracy, I had a real sense of future, of light. As I went to poetry readings, to meetings, to events organized

170

by women, I felt on the verge of a leap. While male politicians seemed bogged down by justification, by the load of responsibility that was the past they had created for their country, women could use their perceived lack of political history to establish possibilities for the future. In 1991, on my return, a theatre director talked of how the end of the Popular Unity years meant 'a leap that stopped mid-air'.[17] In 1991 this leap continued, but this time women were a real part of it.

In August 1991 the Casa de la Mujer, La Morada (established during the dictatorship as a place for meeting and action for women from all over Chile, and organized around cultural and artistic expressions) set up the first feminist radio station in the world. *Radio Tierra* (Radio Earth) aims to 'form a means of communication that is a vehicle for the multiplicity of voices of Chilean and Latin American women'.[18] They aim to 'highlight the protagonistic role of women in society', and their 'challenge is to recreate women in a multiplicity of images that include differences, so demystifying the stereotypes of women formed and used by the mass media.[19] It is crucial here that women identify and assert a 'multiplicity' of roles, actively fighting against the immutability of the presentation of female identity we have studied here. With *Radio Tierra* the women also connect themselves with an indigenous past, looking to ways of learning about a collective past, and aiming to learn from this past in order to 'relate harmoniously with our natural and social surroundings'.[20] Throughout the process they are conscious, as the *arpilleristas* came to be, of an 'apprenticeship', of learning a trade as they create and build, and they have a specific feminist aim:

> This project forms part of a strategy for radio communication for women, with the principal aim of contributing to changing attitudes in society so that men and women see themselves and each other as equals, with similar access to participation in the processes of decision-making at all levels, in this way stimulating active participation in national life.
>
> (Radio de Mujeres 1991: 2)

Throughout, we can see that clear lines of continuity have been established, one is the consistent stating of the gender question, and another is that at the heart of almost all this female activity is the consciousness of creation. This relates to artistic creation, but it also relates to the creation of a consciousness of being a woman within society: not only as wife and mother, roles that they assert as a significant real or potential role in their lives, but also as a worker, as social agent, as citizen. When they have abolished the 'cuestión femenina', and implanted the question of gender they will have accomplished a great part of their objective. But for the moment, they are touching the air, with their feet planted firmly on the ground, and their creative imaginations in flight.

Touching the air is essentially a cultural process, as proven in the first play in Chile to be written, produced, directed and performed entirely by women, where the force of a cultural astuteness was striking. *Cariño Malo* (*Bad Love*) (Stranger 1991) is a set of sketches around a story of a woman who, so betrayed by her failed illusions about love and marriage, and so physically betrayed by her husband, kills him, finding accomplices in two female friends who, it is suggested, may only be representations of different facets of her own personality. *Cariño Malo* was one of the first plays to turn the sights of Chilean theatre inwards, away from social comment and protest, towards a searching of the soul, to a private, hidden, painful, often shameful, and potentially dangerous inner life. Women helped to turn Chilean sight in the theatre to the personal, and the play was a huge success, because the personal needed to be freed in the Chile inherited from Pinochet. The personal had been lost in a quagmire of political reaction, it had been locked away, or used in so many alien ways as subterfuge, as haven, as deposit for fear, an absorber of reality, that the personal realm was tired, and in need of an exploration that did not exploit it more, but that opened it out. Here it was opened out, with *cariño* (tenderness), with humour, with love, truth and daring; and with the explicit aim of forgetting Pinochet. In this way *Cariño Malo* became one of the most significant cultural expressions of these first years of transition. It touched the pulse of the society from which it emerged by searching within the individual, by perceiving the need to recuperate the sense of the openness of the individual space.

There has been a palpable change in the presence of women in Chile. On many levels, the same questions as before remain engrained in society, deep fears still exist about the strength of apparently safe social structures. But, on other levels, the experience of the Pinochet years, throughout which women were at the heart of the struggle against authoritarian rule, is not, as in other situations, a short hiatus in the invisibility of women in Chilean society. This is not a case of women stepping in to fill places left vacant by men. This is a case of positive transgression, a transgression made slowly into worlds of men's power, slowly and with solidarity for men's struggles, a solidarity built on the knowledge that equality and respect between men and women was and is the only progressive way forward. This transgression is built on creation, the creation of bonds, of links, and, most of all, of a multiplicity of languages that articulate, not only female experience, but a history. Women in Chile have become part of the community who feel the air and give memory and experience their words, images and symbols, which, like all words, images and symbols, can belong to everyone.

8

ADJUSTMENT FROM BELOW

Low-income women, time and the triple role in Guayaquil, Ecuador

Caroline O. N. Moser

GENDER BIAS IN STRUCTURAL ADJUSTMENT POLICIES

Widespread concern now exists about deteriorating standards of living, and the severe erosion of the 'human resource base' of the economy, in many Third World countries, after a decade of crisis from debt and recession, and the resulting stabilization and economic structural adjustment policies (SAPs).[1] The fact that the 'social' costs of SAPs have been most heavily carried by the low-income population in both rural and urban areas has resulted in proposals to modify the adjustment process to include 'a Human Face' (UNICEF 1987), with policies to 'strengthen the human resource base' (Demery and Addison 1987).

Despite increasing concern with the plight of poor households, fundamental problems remain. Because of gender bias in macroeconomic policy formulated to reallocate resources, SAPs often have a differential impact *within* households on men and women, and boys and girls (UNICEF 1987; UNICEF n.d.; Elson 1987). Through the analysis of recent research in urban Latin America, the purpose of this chapter is to contribute to the ongoing debate concerning the extent to which SAPs have, even if unintentionally, differentially disadvantaged members of low-income households on the basis of gender. The objective is also to show policy makers the importance of ensuring that current research methodology, such as the SDA survey, shifts from the household as the unit of analysis, towards a more disaggregated approach with greater capability of identifying intra-household differentiation. In addition, the limitations of research which isolate low-income women, outside the context of their households, are identified.

A longitudinal case study of an urban low-income community, Indio Guayas, in Guayaquil, Ecuador, between 1978 and 1988, provides the opportunity to examine the relevance of three kinds of 'male bias', which Elson has identified as underlying many SAPs (Elson 1990: 6).[2]

The first male bias concerns the sexual division of labour, which ignores barriers to labour reallocation in policies designed to switch from non-tradeables to tradeables, by offering incentives to encourage labour-intensive manufactures and crops for exports. Changing household-level employment patterns in Indio Guayas are examined to assess the extent to which any retraining and transfer of labour has occurred, or whether, as Elson has argued, gender barriers to the reallocation of labour have meant greater unemployment for men displaced from non-tradeables, while for any women drawn into export-oriented manufacturing this meant extra work, as factory work is added to the unpaid domestic work which unemployed men remain reluctant to undertake (ibid.; 12).[3]

The second male bias is concerned with the unpaid domestic work necessary for reproducing and maintaining human resources and the extent to which SAPs implicitly assume that these processes, which are carried out unpaid by women, will continue regardless of the way in which resources are reallocated (ibid.: 6). This raises the question as to how far SAPs are successful only at the cost of longer and harder working days for women, forced to increase their labour both within the market and the household. Preoccupation has been expressed regarding the extent to which their labour is infinitely elastic, or whether a breaking point may be reached when women's capacity to reproduce and maintain human resources may collapse (Jolly 1987: 4).

Gender planning methodology, which identifies that women in most low-income households in developing countries have a triple role, provides the necessary tools for identifying how far the issue is elasticity of time, as Elson argues, or changes in the balancing of time (Moser 1986, 1989). 'Women's work' includes not only *reproductive* work (childbearing and childrearing responsibilities) required to guarantee the maintenance and reproduction of the labour force, and *productive* work in income generating activities; in addition it includes *community managing* work, undertaken at a local settlement level. With inadequate state provision of items of collective consumption, and the increasing cutbacks in existing basic services such as water and health services, it is women who take responsibility for the allocation of the limited resources to ensure the survival of their households. Although men are involved in *productive* work, they generally do not have a clearly identified reproductive role. Equally, while they are involved in the community they are generally less involved in the provision of items of collective consumption, but have an important *community politics* role in which they organize at the formal political level, generally within the framework of national politics (Barrig and Fort 1987; Moser 1987).

Because the triple role of women is not recognized, the fact that women, unlike men, are severely constrained by the burden of simultaneously balancing their different roles is frequently ignored by policy makers. In addition the tendency to value only productive work, because of its exchange value, with reproductive and community managing work seen as 'natural' and non-productive, and therefore not valued, has serious consequences for women. It means that the majority, if not all, the work that they do is made invisible and fails to be recognized as work either by men in the community or by those planners whose job it is to assess different needs within low-income communities (Moser 1989).

In examining the impact of SAPs on low-income women in Indio Guayas, the triple role distinction is used for analysing not only the number of hours worked but also, and more importantly, the changes women have made in their allocation of time between work undertaken in the labour market, the community and the household, because of economic crisis. It assists in identifying the extent to which different work is valued by women and men in the community, as well as policy makers, and consequently the extent to which women find that paid work and unpaid work are increasingly competing for their time.

The third male bias concerns the household, as the social institution which is the source of supply of labour, and the assumption that changes in resource allocations in income, food prices and public expenditure, accompanying stabilization and SAPs, affect all members of the household in the same way, because of equal intra-household distribution of resources (Elson 1990: 6). The notion that the household has a 'joint utility' or 'unified family welfare' function is based on the assumption that its concern is to maximize the welfare of all its members, whether through altruism or benevolent dictatorship, and consequently can be treated by planners 'as an individual with a single set of objectives' (ibid.: 26; Evans 1989). In examining the effects of cutbacks in resource allocations on low-income households in Indio Guayas it is necessary to identify intra-household allocations not only of resources, labour and time but also of decision making.

DIFFERENTIATION AMONG LOW-INCOME WOMEN

While the focus of the chapter is on the impact of adjustment processes on low-income women, it is clear that there are severe limitations in studying women in isolation, as is the case in many recent micro-level studies (Commonwealth Secretariat 1990). This has resulted in a tendency to identify income as the basis of differentiation, and to treat as similar the 'plight' of all low-income women. In reality, however, low-income women have not all been equally placed, for not all have been similarly successful in balancing their three roles. In examining differential responses to crisis by women in Indio Guayas it is important to identify the extent to which 'coping strategies' have been influenced by the nature and the composition of the household in which women are positioned. Three determining factors can be identified as likely to affect women: first, the number of persons in the households also involved in productive work and generating a reliable income; second, the particular stage in the household life-cycle when changes occurred; and third, the composition of the household in terms of the number of other females also involved in reproductive work.

The most important factors affecting low-income households during recession and the process of adjustment are clear. In the Latin American urban context three types of changes are identified: first, changes in income, through changes in wages, and levels and sectors of employment for employees, and through changes in product prices and product demand for self-employment; second, changes in consumption patterns through changes in prices in important purchases, especially food; third, changes in the level and composition of public expenditure, through government sectoral spending cuts, particularly in the social sector, resulting in introduction or increase of user charges for services. Despite the difficulties in identifying the degree to which changes are linked to specific adjustment policy, nevertheless, where possible, inferences to causal-relationships are made.[4]

In this chapter Ecuador is briefly described in terms of its problems of debt and recession, and the stabilization and structural adjustment measures implemented by two governments over the past eight years. Data from Indio Guayas provides the basis for a more detailed examination of the impact of recession and SAPs on income, consumption patterns and the level and composition of public expenditure, the three main areas of change identified. The way the situation has changed for women is examined in terms of their different roles, and where there is adequate data, the links to particular policies suggested. Finally, preliminary conclusions are reached relating to the capacity of low-income women to cope with change, with a distinction made between women who are 'coping', those who are 'burnt out', and finally those who are 'hanging on'. A number of policy recommendations are suggested which, while contextually specific, may have wider applicability to other situations in which low-income women have been similarly affected by adjustment processes.

BACKGROUND TO THE RESEARCH

Ecuador and its economy

Ecuador, with a population of 10 million, is one of the smaller Latin American countries, divided into three distinct geographical regions. The coastal plain, with extensive agricultural zones under banana, rice, shrimp and sugar production and cattle rearing, has 50 per cent of

176

the population and is dominated by the export-oriented port city of Guayaquil. The Andean sierra, with 46 per cent of the population, is largely engaged in subsistence agricultural production of mainly potatoes, maize and cereals, and is the location of Quito, the capital city and adminstrative and government centre. Finally, the Amazon Territory, until recently sparsely populated and uncultivated, is the location of oil extraction since drilling began in 1967.

The wider significance of Ecuador lies in the fact that it is a 'median' country. Until 1970 it was one of the poorest in Latin America but, with the discovery of oil, by 1982 it became one of the middle-income countries of the hemisphere, with per capita income equalling its neighbours Colombia and Peru. As a result of its oil-led economic development during the 1970s (it joined OPEC in 1973), the World Bank in 1984 stated that in terms of social progress 'much of the oil income was well spent'. Between 1960 and 1980 more than ten years were added to Ecuadorian life expectancy, death and infant mortality rates dropped by 40 per cent and by 1980 virtually all children attended primary school.[5]

Nevertheless, public sector revenues and expenditures became heavily dependent on oil receipts and external borrowing during the 1970s when public expenditure grew by 9.6 per cent per year in real terms. High oil prices and availability of cheap international credit for oil exporting nations meant that large public sector deficits could be covered, leading to a thirteenfold increase of the public external debt. While the manufacturing industry contributed to growth during the oil boom years, mainly because it was highly protected and catered essentially for the home market, it was neither able to compete in foreign markets nor to cope with the depression associated with the post oil boom, for oil income was involved in current consumption rather than in longer-term productive investments. A strong and appreciating *sucre*, Ecuador's currency, further undermined export oriented manufacturing as well as agriculture. During 1970–86 agriculture grew at less than 2 per cent per annum and while it accounted for 16 per cent of GDP in 1980, by 1988 it was down to less than 7 per cent.

Recession, stabilization and structural adjustment policies

Since 1982 Ecuador has faced a fundamentally changed world economic environment in which new commercial lending has dried up, real interest rates have risen and oil prices have declined. Within the recession period of 1982–8 two particular crisis periods in the Ecuadorian economy can be identified: first, 1982–3 when the bottom dropped out of world oil prices and Ecuador was suddenly plunged into recession, and second, 1986–7 when oil prices fell further and an earthquake caused cut-backs in oil production. During the 1982–8 period two different Ecuadorian governments adopted eight distinct stabilization/adjustment packages.

The first measures taken in 1982 were essentially a stabilization policy to induce adjustment, mainly by controlling demand. A fiscal and monetary programme was introduced, the *sucre* was devalued (its value fell by 17 per cent in 1982, and 32 per cent in 1983) and controls placed on imports, current expenditure and public investment, while the debt payment was revised. Its purpose to reduce real incomes and thereby domestic demand succeeded in that aggregate demand fell to provoke a reduction in GDP in 1982 of 1 per cent rising to 2.8 per cent in 1983. In 1984, a new government intensified demand control measures by a further 30 per cent devaluation of the *sucre*, and suspension of price controls, the reduction in food, energy and fuel subsidies and other measures to reduce the public sector deficit. It also implemented supply side SAPs with incentives to private producers through measures to eliminate export taxes, liberalize imports, and more flexible interest rates, to expand and diversify the export base. One success story has been the development of the shrimp industry which by 1986 had become Ecuador's second largest source of foreign exchange earnings (World Bank 1988).

Despite such measures it proved necessary in late 1985 to devalue the *sucre* once more. When oil prices halved between 1985 and 1986 from US$25 to US$12.7 dollars a barrel, the fall cost the country 950 million dollars, equivalent to 8 per cent of the Internal Product. Following the requirements of the IMF, the government reformulated its SAP in 1986, introducing flexible interest rates and an end to treasury foreign exchange control. However, the public sector deficit continued to increase, rising from 5.1 per cent of GDP in 1986 to 10.5 per cent of GDP in 1987, to 12 per cent in 1988, although this was less to satisfy basic priorities than to meet the political demands of powerful interest groups (UNICEF 1988). By 1986, the economy entered a profound recession with a mere 1.6 per cent growth rate. Further deterioration occurred in 1987, because of both the world global economic situation and the disruption to oil production caused by the March earthquake. By 1988, despite SAPs, Ecuador's economic situation was critical, with the main problems identified as the size of the foreign debt and the acute recession in internal economic activities. Agricultural production for the internal market stagnated together with food processing and textiles as real incomes of the salaried and waged sectors fell. The new government in August 1988 was faced with debt rescheduling, and the necessity to control further the public expenditure deficit.

Barrio Indio Guayas, Guayaquil

Guayaquil is the country's largest city, chief port and major centre of trade and industry. Historically, growth has been linked to the different phases of Ecuador's export oriented economy. As an industrial enclave, its population growth has reflected the national agricultural sector's declining capacity to retain population as much as the city's potential to create industrial employment. It expanded rapidly during the 1970s at the time of the oil boom because of high in-migration rates mainly from surrounding rural areas. This helped swell the population from 500,000 in 1960 to 1.2 million in 1982 and an estimated 2 million in 1988.

Guayaquil's commercial activity is focused around the forty grid-iron blocks of the original Spanish colonial city, which in the 1970s were encircled by the inner city rental tenements. To the north on higher ground are the predominantly middle- and upper-income areas, while to the west and south are tidal swamplands which provide the predominant area for low-income expansion. Settlement of this peripheral zone, known as the *suburbios* (literally suburbs) occurred between 1940 and 1980 when the low-income population excluded from the conventional housing market invaded this municipal-owned swampland.

Research was undertaken in Indio Guayas, the name given to an area of swampland about ten blocks in size, located on the far edge of Cisne Dos, one of the administrative districts of the city. The settlement has no clear physical limits but in 1978 had some 3,000 residents, the majority of whom belonged to the Indio Guayas neigbourhood or *barrio* committee. In 1978 Indio Guayas was a 'pioneer settlement' of young upwardly mobile families, who had moved from inner city rental accommodation and acquired their own 10 by 30 metre plots. They incrementally build their houses of split bamboo and wood on catwalks, relying on irregular water tankers and pirated electricity. During the late 1970s and early 1980s residents mobilized and petitioned local politicians and government to provide infill, drinking water and electricity for their community (Moser 1982: 1).

Data from a household survey undertaken in 1978 showed a mean age for both men and women of 30 years. Free unions were the usual relationship with most households headed by men, resulting in an average household size of 5.8 persons. The community was representative of the lower-paid end of unskilled non-unionized labour. The men were employed as mechanics, construction workers, tailoring out-workers, unskilled factory workers or labourers, while the women were employed as domestic servants, washerwomen, cooks, sellers and dress-makers (Moser 1981). However, the residents of Indio Guayas were an upwardly aspiring community, struggling through hard work and initiative to improve their standard of living and employment prospects for their children, through better health and education.

STRUCTURAL ADJUSTMENT AND ITS IMPACT ON LOW-INCOME HOUSEHOLDS IN INDIO GUAYAS

Changes in income, wages and employment

In examining changes under recession and adjustment, the first fundamental indicator identified is change in income and real resources in cash or kind at the household level. As Cornea *et al.* have identified, 'these resources are needed in the area of food, housing, clothing, transport and to an extent health, water sanitation, education and child care' (Cornea *et al.* 1987: 37). In an urban context changes in household-level income are most directly affected by changes in real wage levels and self-employment rates as well as by shifts in the labour market, both in Indio Guayas and at the national level.

In Ecuador it is estimated that open unemployment increased from 5.7 per cent of the economically active population (EAP) in 1980 to 11 per cent in 1987, accompanied by an 'explosive' growth in underemployment which in 1984 was estimated at 50 per cent for Quito and Guayaquil (UNICEF 1988: 23). This was accompanied by a greater decline in the value of real wages than that of per capita GNP. These changes reflect major features of the IMF 'deflationary' stabilization measures designed to reduce employment in the public sector, to freeze wages through a stringent wage control policy, as well as a high inflation rate especially of food and drink (ibid.). The sectors in which waged work declined particularly severely were those in which the EAP was most concentrated, namely agriculture, industry and construction. In agriculture, participation rates fell from 19 per cent to 6.8 per cent; in industry from 31.5 per cent to 11.5 per cent; and in the construction sector from 50.6 per cent to 28.6 per cent (BCE 1986: 139).

In Guayaquil, the decline in oil revenue and attempts at cut-backs in public sector spending reduced private investment, particularly in the construction sector. Interestingly enough, however, as shown in Table 8.1, in Indio Guayas this did not result in a shift out of the construction sector for men, with numbers working in this sector remaining very much the same, at 11 per cent in 1988 compared to 12 per cent in 1978. However, with a downturn in demand for labour on large infrastructure and office-building projects, most men were employed in the *suburbios* on local house upgrading or the building of extensions. In addition, real wages in the construction sector declined and were worth only half to two-thirds of their 1979 value. Most important of all, changes in contractual conditions occurred, with far fewer skilled men working on fixed-term contracts (with rights to benefits such as social security), and many more increasingly employed, like their unskilled colleagues, on a daily basis. This increased casualization of work in

180

Table 8.1 Breakdown of occupational categories 1978–88 (%)

Census occupational category	Women		Men	
	1978	1988	1978	1988
Professional teacher	1.0	2.9	1.4	0.8
Nursing auxiliary	0.0	2.9	0.0	0.0
Other	0.0	1.9	0.0	1.7
Managers	0.0	2.9	0.0	0.8
Office workers	1.0	1.4	4.6	4.0
Sellers	36.0	33.2	19.7	21.0
Agricultural labour	2.0	0.0	1.8	6.8
Transport	0.0	0.0	8.4	9.0
Artisan/machine operator				
Shoes	0.0	0.0	2.3	0.0
Dressmakers	14.0	17.0	0.0	0.0
Tailors	0.0	0.0	8.8	2.9
Mechanics	0.0	0.0	4.7	6.0
Skilled construction	0.0	0.0	12.0	11.0
Carpentry/lacquer	0.0	0.0	6.0	9.0
Other artisan	0.0	0.0	0.0	2.9
Other factory workers	0.0	2.9	9.8	8.0
Labourers	0.0	0.0	9.8	6.0
Personal services	39.0	35.0	6.1	6.1
Other	7.0	0.0	4.6	4.0
Total	100.0	100.0	100.0	100.0
n = actual numbers in samples	(230)	(131)	(213)	(118)

the construction sector resulted in greater underemployment and unemployment for periods at times up to six months of the year.

The number of male sellers remained the same at around 20 per cent. The number of artisans also showed stability, although more shoe-makers and tailors were evident in 1978 than in 1988, while slightly more mechanics and carpenters were found in 1988. A breakdown of factory and labouring work shows the only real area of expansion for male employment was in sea products, particularly the shrimp industry. Around 7 per cent of male heads of household had jobs in this well-paid sector. With shrimp farms covering an extensive area along the Pacific coast this resulted in a small but growing male circulatory migration pattern from Indio Guayas in which only one weekend in three was spent at home. For the men in Indio Guayas, therefore, few opportunities to shift from non-tradeables to tradeables in labour intensive manufacturing materialized with SAPs. Consequently, it is export crops, such as shrimps, biased in favour of male employment, which are proving increasingly attractive for urban men, regardless of the consequences for social relations in urban households.

Women in Indio Guayas were severely affected by the fact that fewer men than before were generating a reliable income, and the declining value of men's wage packets. First and foremost, more women had to work, with female participation rates increasing from 40 per cent in 1978 to 52 per cent in 1988.[6] Although more women were working, the majority were not able to take up new opportunities in the labour market. No new factory work recruiting female labour was created, and women were forced to remain in areas of domestic service and street selling in which they always worked. Although gender segregation in the labour market protected women's employment in the service sector, this has been achieved only at lower rates of pay. As shown in Table 8.1, in 1978, 39 per cent of economically active women were domestic servants, cooks or washerwomen, while by 1988, 35 per cent were in the same occupations, although average wages fell to two-thirds of their 1979 value. The same was true of street and front-room selling, probably the most involutionary part of the informal sector given its capacity to absorb additional labour without increased productivity. In 1978, 36 per cent of women were street sellers, as compared to 33.2 per cent in 1988, with the numbers of dressmakers remaining the same at around 5 per cent.

The biggest change was the increase of those in professional (teachers and nursing auxiliaries) and office jobs, growing from 2 per cent in 1978 to 10.1 per cent in 1988. Comprising mainly younger women, this group reflected the slowly increasing socio-economic differentiation of Indio Guayas as the city's spatial expansion made the area more attractive to higher-income groups.[7] Data from the 1988 survey shows that of those households with working daughters, the largest number were employed as shop workers (36 per cent). Although the majority of them had completed secondary school, they were either on short-term contracts or selling on commission. A further 15 per cent were in professional and office jobs, and over 20 per cent were in domestic service.

Elson's concern about gender barriers in the reallocation of *paid* labour did not emerge as a major issue in Indio Guayas. What was more

Table 8.2 Changes in the number of the economically active within the household in Indio Guayas 1978–88 (%)

Number working	1978	1988
1	49	34
2	32	32
3	11	20
4	6	11
5 or more	2	3
Total	100	100

important than gender barriers was the paucity of new jobs in tradeables for *anyone* within the community. No new jobs have been generated for women, and only a few for men, and these are in Guayaquil city itself.

A second fundamental change resulting from changes in income was changes in the number of the economically active in the households. As shown in Table 8.2 fewer households depended on the income of one earner, with a decline in the number of households with one working (from 49 per cent to 34 per cent) and a significant increase in the numbers with three or more working (from 19 per cent to 32 per cent). With more people in the households having to work, the strategy of who worked and who stayed at home depended on both the ages of the children, and the number of daughters who replace their mothers in undertaking reproductive tasks in the home.

A third change was in the composition of households with the number of woman-headed households increasing from 12 per cent to 19 per cent in the decade 1978–88. While there were more older widowed or deserted women in the community than ten years previously, there was also an increasing number of *de facto* woman-headed households, as shown in Table 8.3. In some cases men temporarily migrated to the rural areas to work in agriculture or the shrimp industry, leaving their family in Indio Guayas because of better opportunities for female employment and children's education in the city. Frequently, however, these men soon set up other households in the rural areas, resulting in a declining responsibility to the urban household.

Table 8.3 Changes in the composition of the household in Indio Guayas 1978–88 (%)

Household type	1978	1988
Nuclear	52	51
Woman-headed	12	18.3
Extended	25	25.7
Other	11	5
Total	100	100

Although the number of nuclear and extended families has remained fairly constant (52 per cent and 25 per cent respectively), the composition of extended families changed. Fewer had dependent grandparents, while more had married sons and daughters and their income earning partners. Young couples were less likely to leave the family home to start their own home than they were ten years previously. Two reasons can be put forward to explain this trend: first, households were more dependent on additional income earners; second, the incentives to replicate the 'pioneer process' experienced by their parents – who in the

1970s invaded mangrove swamps and squatted without water, electricity, roads or indeed land rights – are not as great in the current politico-economic context.[8] The option of land invasion and settlement consolidation in the 1970s faced the new reality of bankrupt local and national governments and public utilities which no longer had the financial capacity to expand already over-extended services into more far-flung squatter settlements.

Changes in consumption patterns

During the decade 1978–88 the cost of living increased for low-income households. One of the principal long-term root causes was the agricultural development policy adopted during the petrol boom period when changes from subsistence basic grains production to export oriented cash crops provoked a decline in foodstuff production for the home market. Inflationary rises in food prices were also caused by the removal of subsidies on basic staples and the *sucre*'s devaluations, which increased the prices of imported foods. With wages frozen, while food prices and other essential items such as energy and petrol rose in price (with knock-on effects for transport costs), real incomes fell. It has been estimated that in 1984 the minimum salary for those with stable wage employment met only 65 per cent of the costs of the family shopping basket. Futhermore it was also calculated that the purchasing power of informal sector earnings covered only 35 per cent of the value of family food requirements (UNICEF 1988: 3). Price increases in staple food items in Guayaquil during the one-year period from June 1987 to June 1988, for instance, ranged from 50 per cent increase on milk to 194 per cent on potatoes. Other important changes included 79 per cent on eggs, 55 per cent on fish, 117 per cent on tomatoes, 25 per cent on plantain and 93 per cent on rice.

This resulted in changes in household diet. Sub-sample data for the period July 1987 to July 1988 showed, for instance, that 42 per cent of households no longer drank milk at all, while in those where milk was still consumed average consumption decreased sharply from 4.6 litres per household per week to 1.4 litres. (The corner shop which used to order 36 litres of milk per day now orders only 12.) Similarly the average number of times per week fish was eaten declined from nearly three to under two, for eggs from an average of 4.9 per week to 2.63. Potatoes gave way to plantain and fresh fruit juice to powdered fruit drink or water. Detailed data on daily menus shows the following: first, a tendency to eat smaller quantities per meal and thereby stretch the food previously eaten at one meal to cover both mid-day and evening meals; second, to eat fewer meals, with first supper and then breakfast cut out. In July 1988 one-quarter of households ate one meal a day with the

184

breakfast cut at times, justified on the misguided assumption that 'the children get free milk at school'.

It was also harder to pay for essentials. High inflation during the previous years meant that many local small shops which previously granted credit were unable to do so because of losses incurred by bad debts. At the same time the decline in stable household income resulted in a reduction in monthly or weekly bulk purchases in the centre. Non-perishable food items such as cooking oil, lentils, spaghetti and sugar were more likely to be bought in small quantities at higher prices at local shops. Another indicator of change in consumption patterns was provided by the number of broken-down refrigerators which were bought during the boom period between 1978 and 1982, and which households were subsequently unable to mend. The lack of a refrigerator in a hot climate means that food has to be prepared on a daily basis, and this has consequences in time spent cooking. Such a pattern reverses the trend over recent years when women saw their cooking time reduced as they changed from cooking on kerosene or charcoal to gas.

Changes in dietary patterns resulted in a deterioration in the nutritional status of households, particularly of children. A two-week survey of all children under 12 attending a local health centre revealed that 79 per cent were suffering some level of malnutrition, of which 28 per cent were first degree, 38 per cent were second degree and 13 per cent were third degree. Although this data is not sufficient to indicate a trend, it nevertheless reflects the findings of national-level statistics on nutritional status (see UNICEF 1988; Freire 1985, 1988). It was also evident (from data collected at a qualitative level) that women fed themselves last, and ate least, with anaemia a much-voiced complaint.

In a situation of personal debt, loans from wealthier kin were increasingly sought after, while the sale or pawning of jewellery or other valuables occurred more commonly as did borrowing smaller amounts from neighbouring women. A number of community-level self-help saving schemes were developed by women within local *barrio* committees, where a small amount was put away each week in a savings club to provide a small capital source to help pay school matriculation costs or for new clothes at Christmas. The *cadena* was another initiative introduced in the past five years in which forty families paid weekly quotas into a rotating fund used exclusively for the purchase of housing materials. Such self-help saving schemes are again mainly organized by women.

Changes in levels and composition of public expenditure due to changes in governmental sectoral spending

During the oil boom social sector spending benefited considerably. While the total national budget grew at an average of 5.8 per cent during the 1972–82 period, the social sector budget grew on average by 11.3 per cent per annum. The social sector share of the national budget rose from 31.7 per cent in 1972 to 39.5 per cent in 1980. With the 1981 adjustment measures it fell to 30.5 per cent by 1985, picking up to 34.2 per cent in 1986. The highest share of the national budget is on education (26.5 per cent), followed by health (6.5 per cent) and social welfare (1 per cent) (World Bank 1988: 97–8). However, these figures hide per capita shifts and UNICEF has argued that while the health budget represented 7 per cent of the national budget between 1979 and 1984, its real value per capita declined by 18 per cent between 1981 and 1985 (Suarez 1987: 49). In an urban community such as Indio Guayas, the problem with social services related not so much to lack of provision as to the quality of service provided, and the extent to which it was free or purchased. Payment can either be direct, through private sector provision, or indirect, through hidden transfer charges for users in state provision, or through unpaid labour time in non-governmental organization service provision.

In 1978 in Indio Guayas, the nearest health services were provided by the out-patient departments of the government Hospital Jesús de María, and the USAID-supported Hospital de Guayaquil on the edge of the suburbios, together with other city centre private services. The decade 1978–88 witnessed a proliferation in the growth of private practices in the sector itself, a consequence as much of the oversupply of professionals as the demand for their services. For instance, within seventeen blocks on the main access road which passes through Indio Guayas there were six general practitioners, two surgeons, three paediatric specialists, four dentists and an obstetrician as well as three pharmacies and a laboratory. The most important state provision was a highly overcrowded Primary Health Centre located a bus ride away, while the only surviving NGO provision is a small health clinic, financed by a foreign religious order, which provided a doctor's surgery and drugs at a subsidized rate.

While data on local health sector spending cuts were unobtainable, there was a general consensus indicated in the survey that government-run services had deteriorated. Asked which service they would use if a member of their family fell ill, 50 per cent responded that they would go to the state hospital, 35 per cent to a private doctor and 11 per cent to the health centre. Asked where they had gone the last time a member of the family had been ill, 42 per cent said they had used the hospital, 48 per cent a private doctor while only 9 per cent had used the health centre. In the 1987–8 period the doctor at the subsidized NGO clinic said that the number of patients had doubled because patients could no longer afford to pay for private services. The picture of healthcare which emerges is therefore one of increasing privatization with costs

186

borne by the low-income population, and the failure of most subsidized programmes, such as the UNICEF-supported Primary Health Care Programme, to survive in the long term (Moser and Sollis, 1989).

Ten years ago the only educational facility in Indio Guayas was a cane-walled primary school with seventy to eighty children per class. By 1988 the same school was a large complex built of cement and bricks, operating a three-shift schooling system, with financial assistance from Plan International, an international NGO. The UNICEF Pre-School Programme survived in Indio Guayas during 1982–8 largely because of extensive community-level support, which included the building of a community centre. Local members were nominated as teachers and paid a 'stipend' for their part-time work. The state primary school, however, was similar to those at the national level, with inadequate equipment, low quality of teaching and outdated teaching methods. At the national level, primary schooling extended to 86 per cent of the population, but only 35.7 per cent were in the appropriate grade for their age. Despite such limitations, increasing competition to enter the labour market led to an expansion in demand for formal educational qualifications. In Indio Guayas there was widespread concern to educate both sons and daughters for as long as possible. Low-income families invested considerable proportions of their income and savings to educate their children. The survey showed that in 1988, 36 per cent of households sent up to five of their children to private fee-paying schools (for a year or more), in the expectation that it would provide better education than the equivalent state school. While a quarter had gone to pre-primary schools, the highest investment was made at primary level, attended by 65 per cent, with only 12 per cent at secondary level.

Although state education was technically 'free', in reality numerous charges were transferred to the parents. Over the decade 1978–88, the annual school entry 'matriculation' fee, particularly at secondary level, became a crippling expense for households unable to save during the year. Matriculation fees, set by individual school head teachers, together with their executive councils, varied according to the status of the school but cost up to or more than one month's salary of a domestic servant. When additional costs such as uniform, school books and transportation were included, the annual cost of secondary school per year reached between one and two minimum monthly salaries. One, more prosperous, parent in Indio Guayas who paid 2,400 *sucre* matriculation fee in 1983 paid 10,500 *sucres* in 1988, an increase of 337 per cent. The minimum salary in August 1988 was 19,000 *sucres* per month, while the total year's schooling costs were calculated by the same parent as likely to be 40,500 *sucres*.

The combined effect of a fall in real income due to cuts in wages, inflationary food prices and increased expenses on education and health meant that the majority of households in Indio Guayas were poorer in real terms than in 1978 with a marked deterioration in 1986–8. The extent to which this affected all members of the household obviously related to who within the household made up the shortfall. With men contributing less or, at times, no cash to the household budget, the crucial question is whether women had to find more resources. This is most usefully analysed in terms of their triple role.

THE IMPACT OF RECESSION AND ADJUSTMENT PROCESSES ON WOMEN IN INDIO GUAYAS

The productive role of women

The changes outlined above had a number of important implications for women in their productive role. More women were working in income generating activities than were doing ten years previously, in many cases as the primary, reliable income earner. Although badly paid, both domestic service and selling provided more stable and reliable work contracts than the increasingly irregular day labour available for men.

The main factor determining which women were working was the number of persons in the household involved in productive work and generating a reliable income. Some women had always worked. Four-fifths of women in the sub-sample worked during the 1978–88 period; of those, nearly half (48.14 per cent) worked throughout the period. Reasons given included: because as heads of households they always were the sole income earner; because the low income of their spouse meant the household depended on their income; or because they chose to work. The particular stage in the household life-cycle was a second factor which also affected which women worked. Those few women not working tended to be in extended or women-headed households where daughters, sons or in-laws contributed to household income and the woman's duties were entirely confined to reproductive work relating to household needs. This option also suggests that the size of household affected its level of income.

Whether women went out to work depended on factors external and internal to the household. The sub-sample shows that 18.5 per cent entered work during the 1982–3 period, with the same number entering in the 1987–8 period. All stated that they had started working because their household income was insufficient to cover family needs, with '*no alcanza*' (we haven't enough) a commonly heard phrase. The certainty that their entry coincided with the two crisis periods when macro-economic measures resulted in declining wages and increased food prices suggests a direct causal link. In particular, inflationary food prices in 1987–8 were identified by women as a direct reason for going out to

188

work. Within the household the other important reason identified was the cost of secondary-level schooling, again exacerbated by inflationary matriculation fee increases in the past two years. Households experiencing additional pressure were those with one or two children in secondary schooling.

Where women worked depended not only on their skills but also on levels of mobility. Least mobile were those working from home because of young children or their husband's dislike of wives' employment. The less skilled ran highly competitive front-room shops, while the more skilled worked as dressmakers. Slightly more mobile were those selling cooked food on the corner of nearby main roads, leaving their children for short periods of time. Laundry women had to be able to travel for periods of the day, while the most mobile were domestic servants, many of whom left the house at 6 am in order to cross the city, returning at 8 or 9 pm at night. Evidence suggests that among unskilled women there was a correlation between mobility and the amount earned.

It is clear that women were likely to be making up the shortfall in household income. Not only were more involved in income earning activities than ten years previously, but with the decline in real wages they were also working longer hours. This was most evident in the case of poorer, less skilled women in laundry or domestic service, particularly heads of households, who always worked. Whereas in 1978 they 'did' for one family or possibly two, the majority now fitted in two to three families, working as long as sixty hours a week, including Saturdays, in order to earn the same amount in real terms as before.

In addition, women were forced to seek work when their children were younger than before. Even the poorest women were reluctant to go out to work until their children reached primary school age, while among higher-income families secondary school entry was preferred. The evidence suggests that most women now worked once their children were in primary school, with increasing numbers entering the labour force as soon as possible after their last (intended) child was born. Of those in the sub-sample entering the labour market since 1978, all entered prior to their youngest child's tenth birthday, and nearly 80 per cent before their sixth birthday.

A third factor which affected women's ability to work was the composition of households in terms of the number of other females involved in reproductive work. Although the presence of other adult women able to assist in reproductive activities was sometimes important, it was the number of daughters, and their ages, which most directly determined strategies followed. With a greater number of women both working, and working longer hours, daughters were forced increasingly to dovetail their schooling to their mother's working hours. Although the half-day school shift system made it possible for daughters to

189

continue at school while taking on reproductive responsibilities, it nevertheless meant they had less time for school homework than did their brothers. As a result, girls were often disadvantaged in terms of academic achievement, causing them to fail in school. The future productive potential of daughters, therefore, was increasingly constrained by their reproductive activities.

The reproductive role of women

In their reproductive role as wives and mothers, women were affected by the adjustment process, above all in terms of conflicting demands on their time. Despite the fact that more women were working in 1988 than in 1978 the cultural norm in Indio Guayas that reproductive work is women's work did not change, as men did not take on new reproductive responsibilities (other than in isolated examples, particularly of house-hold enterprises such as tailoring where men assisted in childcare and cooking while women did the daily travelling to the subcontractor). Where Elson's concerns about gender barriers are justified, therefore, is in women's unpaid labour, since men did not take on more domestic unpaid labour. Increasing pressure for women to earn an income resulted in less time than before to dedicate to childcare and domestic responsibilities. The sub-sample showed that the average number of children per household was 4.78, indicating a substantial number of years when close access to the home was necessary.

The capacity of women to balance productive and reproductive work depended on both the composition of the household in terms of other females, and on the particular stage in the household life-cycle. Women with only very young children, forced out to work, had no alternative other than to lock them up while away, obviously the most dangerous solution. The eldest daughter very rapidly assumed responsibility for her siblings but was not put in charge of cooking until the age of 10 or 11. In this situation women started their day at 4 or 5 am, cooking food to leave ready for their children to eat during the day, and doing additional domestic tasks on their return. Once daughters were able to undertake cooking as well as childcare responsibilities, women did not get more rest, but worked longer hours outside the home. Those households with more than two daughters made maximum use of the half-day school shift system by sending out different daughters to different shifts, thus freeing the mother for full-time work.

In the sub-sample there were a small but growing number of households effectively headed by daughters who undertook not only all reproductive activities but also attended community meetings on Saturdays and Sundays, thus also fulfilling their mother's community managing role. Despite the fact that women in this situation still had a

190

number of young children, their only real role was a productive one. In 1978 this phenomenon was not apparent, suggesting that women were better at balancing their reproductive and productive activities. The situation was exacerbated in those households headed by women, and became particularly problematic in those households where men migrated for work reasons. For instance, although only 6 per cent of the male labour force had migrated to work in the shrimp industry, this initiative was the direct result of structural adjustment incentives. Increasing employment opportunities in this rural export oriented sector, while increasing male income had indirect costs such as the breakdown of some marital relationships, and a reduction in male economic responsibilty for the household.

A reduction in women's time for reproductive activities had a number of important implications for their children. First, young chidren received less care than before, when locked up or unattended, or attended by elder sisters. They were more likely to play truant from school, and to become street children, although not necessarily identified as such, roaming locally, running errands in return for food, and protected by sympathetic neighbours only as long as they remained in the street. Second, young children were suffering additional nutritional problems when not fed by their mothers. Food left for division amongst children was often not fairly divided, and there were often nutritional problems in food cooked by elder siblings. Third, elder daughters forced into reproductive activities at a young age were themselves suffering from less parental care and guidance. Although socialized to assist their mothers with domestic tasks, daughters did not automatically accept responsibilites thrust on them. Resistance could cause conflict with their mothers.

Fourth, a reduction in parental control was detrimental for sons. One of the greatest concerns expressed by women forced out to work was the fact that it reduced their capacity to control teenage sons, who were more than tempted to drop out of school, become involved in street gangs and be exposed to drugs. This problem was exacerbated when the parental responsibility of the father also ceased, for instance due to out-migration. In Indio Guayas it was felt very strongly that the number of street gangs and associated theft had increased considerably during the previous five years, with the level of drug addiction, especially from cocaine paste, now a widely perceived problem. The doctor at the local health clinic confirmed that over the past two years men had increasingly been seeking guidance about drug addiction.

Women in their reproductive role had responsibility for household budgeting and ensuring not only that sufficient food was provided daily but also that larger bills such as school costs were paid. However, their control over the allocation of total household income was very limited.

The data from the sub-sample showed that joint decision making or sharing of resources within the household was not common, with most women receiving a daily allowance from their partners. Although in most cases the amount received had increased during the previous year this increase had not kept pace with inflation. Men did not necessarily allocate increased income to household expenses, such that the notion that the household had a 'joint utility' or 'unified family welfare' function was not borne out in reality.

Supportive evidence is provided by the clear causal linkage which can be drawn between changes in consumption patterns and increased domestic violence within the household. In the sub-sample, 18 per cent of women said there had been a decrease in domestic violence. These were mainly women earning a reliable income, who identified more respect from their male partners as associated with their greater economic independence. While 27 per cent said nothing had changed, 48 per cent said there had been an increase in domestic violence, identifying this as the direct consequence of lack of sufficient cash, and stating that it always occurred when the woman had to ask for more money, in other words when attempting to control how resources were pooled. A distinction was made between those men who became angry out of frustration from not earning enough and those who became angry because they wanted to retain what they did earn for their personal expenditure, identified by wives and partners as other women and alcohol. In both cases, however, the consequence was the same, with men beating women. Additional problems emerged with increasing drug addiction, which unlike alcohol was consumed mainly by younger men. While undertaking fieldwork, the first suicide known to the community leader occurred when a young male cocaine addict killed himself after a confrontation with his wife over the allocation of most of his income to service his addiction rather than feed his three young children.

Comparative anthropological research shows that the pressure to earn an income made it increasingly important for women to control their fertility effectively. Although the stage in the household life-cycle is an important determinant, nevertheless the 1988 sub-sample showed that while a quarter of women were not using any form of contraception, and a further quarter were using the coil, 42 per cent had undergone tubectomy at the birth of their youngest child. However, women were not in control of this fertility decision, as their husband's permission was required for the operation.[9]

The community managing role of women

During the late 1970s and early 1980s women played an important community managing role in struggling to acquire infrastructural resources such as infill, water and electricity for the area (Moser 1987). This centred around popular mobilization, linked to particular political patronage with intensive activity at election time. Cut-backs in public spending after 1983 meant that patronage of this type virtually ceased by 1988. In this context NGOs played an increasingly important role in service delivery, not only as in the UNICEF education and health programmes but also, for instance, in community-level 'developmental' programmes introduced by Plan International since 1983.

This has had important implications for women who spent more time than before in a community managing role, in order to negotiate NGO participatory delivery systems. These programmes, such as Plan International, were based on the voluntary unpaid involvement of women on a regular long-term basis. Women community leaders were expected to provide access to the community, and, along with paid community development workers, to supervise the allocation of resources for development programmes. In order to get access to resources families were required to attend weekly meetings and undertake community-level voluntary activities. Other than in leadership roles, participation was almost entirely by women. As an extension of their domestic role, women took primary responsibility for the success of community-level projects. In some cases, as in the UNICEF Pre-School Programme, these were managed by the community, while in other cases top-down provision resulted in participation only in the implementation phase.

The decreasing provision of services by the government led women to recognize the importance of encouraging the entry, and ensuring the long-term survival, of community-based programmes. Since it was the lack of time which often resulted in failure, women were forced to find the time (to ensure NGOs continued working with them). In August 1988, for instance, up to 200 women met for three hours every Sunday afternoon in order to get access to Plan International's community-level housing improvement programme. At the same time, women with constantly sick children made it their business to attend the Saturday afternoon health talks run by the NGO health clinic. Even men were gradually beginning to recognize this role. A local carpenter summed it up when he said 'I earn the money, and my wife looks after the children and attends the meetings.'

CONCLUSION: WOMEN, TIME AND THE TRIPLE ROLE

Policy makers have become preoccupied with the idea that recession and adjustment have resulted in an extension of the working day of low-income women. The evidence from Indio Guayas shows that the real problem is not the length of time women work but the way in which they balance their time between activities undertaken in their reproductive, productive and community managing roles. In addition it has become important to differentiate such women's work inter-generationally, so as to identify when the extra time comes from daughters rather than mothers. In both cases the extra demands on time are demands on women, with the number of women involved increasing considerably.

Over the decade 1978–88, low-income women in Indio Guayas always worked between twelve and eighteen hours per day, depending on such factors as the composition of household, the time of year, and their skills. Therefore, the hours worked did not change fundamentally. What did change was the time allocated to the different activities undertaken by them. The necessity to get access to resources forced women to allocate increasing time to productive and community managing activities, at the expense of reproductive activities, which in many cases became a secondary priority. The fact that paid work and unpaid work were competing for women's time had important impacts on children, on women themselves and on the disintegration of the household.

It is clear that in the case of Indio Guayas, gender bias in macro-economic policy formulated to reallocate resources had differentially disadvantaged low-income women. Not all women could cope under crisis and it is necessary to stop romanticizing about their infinite capacity to do so. At the same time, they did not form a homogeneous group, and, in terms of their capacity to balance their three roles in the changing situation, fall roughly into three groups.[10]

The first group, women who are 'coping', are those women balancing their three roles. They are more likely to be in stable relationships, with partners who have reliable sources of income. The household income is likely to be supplemented by others working, and there may be other females also involved in reproductive work. About 30 per cent of women are coping.

The second group are women who are 'burnt out'. These are no longer balancing their three roles, and their productive role predominates. They are most likely to be women who head households or are the primary income earners, working in domestic service, with partners who make no financial contribution to the household. They are often older women at the end of their reproductive cycle, physically and mentally exhausted after years of responsibility for a large number of dependants. Their inability to balance their roles results in a tendency to hand over all reproductive responsibilities to older daughters who cannot or will not take all the necessary responsibility. The consequence is that their younger, still dependent children drop out of school, and roam the streets. About 15 per cent of women are no longer coping, are already casualties and burnt out.

194

The third group, women who are 'hanging on', are those who are under pressure but still trying to balance their three roles, making choices depending on the composition of the household and the extent to which other household members are providing reliable income. Some are women without partners, who, if they are the main income earners, have sufficient support from other females. Others are women with partners who have been forced out to work to help pay for the increased household expenses. These women are using up future resources in order to survive today, sending their sons out to work, or keeping their daughters at home to take over domestic responsibilities. About 55 per cent of women are invisibly at risk, only just hanging on.

Ultimately, only the introduction of a gender-aware planning perspective will change current policy approaches to low-income women such as these. Nevertheless, there are also a number of pragmatic changes which can assist them in the short term. Although these women identified as 'burnt out' are an obvious target group for assistance, those identified as 'hanging on', although less visible, are a greater priority in terms of policy prescriptions for human resource development in Indio Guayas. Athough policy makers may not prioritize the problems of these women, account must also be taken of the fact that their daughters are often losing schooling. It is essential, therefore, to ensure that compensatory programmes, designed to 'protect the basic living standards, health and nutrition', target not only the 'burnt out', but also those 'hanging on' (UNICEF 1987: 134).

Too much focus on women as 'victims' of adjustment may undermine their capacity to organize themselves within existing community-level structures and discourage bottom-up, self-help solutions. 'Meso' and sectoral programmes and projects whose priorities include 'expenditures and activities which help maintain the incomes of the poor' (ibid.) are more likely to succeed when they are planned in collaboration with, and implemented by, community-level organizations, especially those led and organized by women.

For, although women are 'victims' of adjustment they are also a largely untapped resource with their community managing role unrecognized. Yet they are prepared to invest commitment and time in those interventions directly or indirectly likely to benefit their families and children. Local women know community needs, and can identify the particular constraints, often much better than professionals who often neither know the communities, nor have the same level of commitment to project success.

All over the world low-income women are providing voluntary labour in their community managing role, in their productive role, are working below the minimum wage (often for no more than a stipend), at the

195

9

'PEOPLE HAVE TO RISE UP – LIKE THE GREAT WOMEN FIGHTERS'

The state and peasant women in Peru

Sarah A. Radcliffe

INTRODUCTION

Over the past ten years peasant women, or *campesinas*, have entered the mainstream of political protest in Peru by participating in land invasions and demonstrations, by forming local defence and income-generating groups, and by making representations to government. All of these political activities now bear the stamp of women's gender-specific concerns and ways of organizing (Radcliffe 1990a, forthcoming; Andreas 1985), as do urban social movements (Blondet 1987; Radcliffe 1988; Barrig 1986; Hernández 1986; Vargas 1986). *Campesina* women are also active in negotiating their gender identities, their femininity, in the sphere of political culture[1] (see Bourque 1989). The forging of these identities, from within the particular spatial and historical context of Peru in the 1970s and 1980s, forms the subject of this chapter.

Women in the Peruvian peasant unions demonstrate a distinct femininity or female identity compared with that represented in Hispanic, Catholic urban culture, which in itself is differentiated between popular elements and the elite and middle class. Such distinct femininities arise through the different class, ethnic and regional cultures in which women live and act, and in turn are negotiated and changed in relation to the state's actions (Radcliffe 1990a).[2] The state,[3] through its policies, programmes and rhetoric, sanctions and promotes certain femininities, while marginalizing or suppressing others. This project is never complete, however, as peasant women (and others) resist and challenge the state's notions of appropriate female identities and actions, and the state's impacts are spatially uneven (especially in Third World countries such as Peru). Nevertheless, peasant women express and perceive their (gendered) political priorities in relation to a state which is configured around a particular gender-ethnic-class group and power structure (see Alvarez 1990: 22), and take on board some aspects

197

of the state's gendered images as components in their own identity. An analysis of the state is therefore crucial to an understanding of gendered identities (masculinities and femininities) in political activism and formation of agendas, not only in peasant groups.[4]

Focusing on the 1968–90 period, this chapter examines in a preliminary way how the relations between socially articulated peasant femininities and state-sponsored images of women were negotiated, and through which practices. In the arena of political culture and political symbolism, peasant women protested against attempts to shape their identities. Protest in this sense refers not to classic actions of formal protest (strikes, demonstrations, and so on) but to resistance by ethnic minority peasant women to attempts by the state to hegemonize female identities (see Laclau and Mouffe 1985; Mouffe 1979).

The chapter starts with a brief overview of the relations between the state and peasantries through time, followed by discussion of the theories on gender and the state. Next, the case of Peru illustrates the nature of state interventions in this sphere over three periods. The actions of women and men within the two major peasant unions in Peru to retain and elaborate their own gender regime[5] are then described, highlighting their negotiation with the state over the political culture of representations and practices fundamental to gender identity. The detailed example of Puno department provides insights into the way that femininities can vary spatially.

PERU – THE STATE AND PEASANTRIES

Since its conquest by the Spaniards in 1532, Peru has been a highly spatially and socially differentiated society. For a long time centrally organized from the capital, Lima, which acts as the focus of development, politics and high culture, Peru's official priorities have been urban oriented and 'European'. The influence of the traditional Catholic Church has been, and continues to be, most marked in urban culture, contrasting with the syncretic religion characteristic of rural and peasant areas, and with the recent development of Liberation Theology in both rural and urban communities.

Interaction of governments with peasantries through history has been characterized by the poor treatment of the latter and intermarriage of populations, resulting in a large 'mixed-race' *mestizo* group. During the twentieth century, the state has been active in the promotion of a *mestizo* nation, one based on progressive 'Westernizing' and de-indigenizing cultural change, in which the 'culture of progress' (Harvey 1987) has systematically denied the validity or concerns of indigenous groups (Portocarrero 1988; Radcliffe 1990b). Economic nationalism too has been rooted in Western models of economic growth, in large part dependent capitalism. Peru's radical military 'experiment' of the early 1970s defined its 'Third Road' as lying between capitalism and socialism (Slater 1989; McClintock and Lowenthal 1983).

As a consequence of these factors, peasants have played a relatively minor role within the society and economy of Peru, while representing one of its most disadvantaged groups. Concentrated in the Andean Highlands, independent peasant communities comprise around 15 per cent of the rural population, although they control 29 per cent of agricultural land, most of it poor-quality pastures and fallow land. From 1987–90, GDP declined in Peru by some 23 per cent, with devastating effects on real incomes, which fell from a factor of 100 in 1980 to 43.9 in 1990. By the mid-1980s the independent peasantries' share of national income reached only 3 per cent (Gonzales 1984: 17). Peasant labour represents around a fifth of the national total, although they are responsible for over a quarter of gross agricultural production (ibid.).

To protect their livelihoods, peasants have been active in political spheres as varied as strikes, marches, unions and federations, land invasions, ignoring estate-owners' strictures, migration and rebellions. From the 1940s, national peasant unions developed which were recognized, but not controlled, by the state. Today, the major unions include the Confederación Campesina del Perú (CCP: Peruvian Peasants Confederation, founded in 1947) and the Confederación Nacional Agraria (CNA: National Agrarian Confederation, founded in 1974). Representing peasants and acting as pressure groups over government policies on pricing, rural wages and land titles, these confederations have increasingly coordinated their actions in recent years (Monge 1989; Eguren 1988). The place of peasant women within these confederations provides this chapter with illustrations of the mutual influence of state and civil femininities.

Andean *campesina* women have suffered from a 'triple oppression' (Bronstein 1982) within Peru's history, as female peasants in a Third World country. With high rates of illiteracy and mortality, and misrecognition (at local and national levels) of their participation in agricultural work, *campesinas* have consistently been marginalized from state concerns. Moreover, their role in peasant communities has been silenced by an indigenous masculinity which claims a 'public' political voice and authority embedded in popular rural culture (see Bourque and Warren 1981). It is within this context that the construction of gender identities of *campesina* women takes place.

GENDER, IDENTITIES AND THE STATE

It is now generally agreed that gendered power relationships vary, according to the mode of production and cultural-historical context

(Alvarez 1990; Connell 1987; Scott 1986; Anthias and Yuval-Davis 1983; Afshar 1987). Not only do the practices of power relations vary, but so do the meanings ascribed to these relations. According to this analysis, different power positions are available to men and women in discourse and practice, by which they negotiate and confirm their interests (Holloway 1984). Nevertheless, some theories on gender-differentiated positions tend to reduce these insights to a male–female dichotomy, rather than looking at the variety of subject positions available to women (and men) resulting from the existence of multiple coterminous femininities (and masculinities) (cf. Anthias and Yuval-Davis 1989; Jackson 1991).

As discussed here, masculinities and femininities refer to the gendered identities created through everyday practices and discourses. Historically constituted, they are the way in which male and female bodies behave to demonstrate their gender identification. Masculinities and femininities are multiple, and give representation and self-repre-sentation to underlying sexual difference, without being directly related to that biological opposition. Masculinities and femininities thus have a socially-mediated relationship to male and female bodies (Cocks 1989; Fraser 1989). As defined here, the identities of femininities and mascu-linities are both historical and in constant flux, changing with changes in social practices. They become geographically and ethnically placed, although their components may be shared by different groups. Most importantly, masculinities and femininities can overlap in space, inter-acting and thereby defining themselves in relation to other gender identities present in the same location.

Socially constituted femininities and masculinities are significant not only because they locate and identify male and female persons, but because they give powers to act in various political contexts. This power to act exists in everyday behaviours and practices, as well as in formal political spheres traditionally associated with 'politics'. Moreover, these gender identities enter into state institutions, which are constructed around masculinities and femininities and intervene in order to shape and reward specific masculine and feminine identities (Anthias and Yuval-Davis 1989: 4–6).

So how does this notion of masculinities and femininities help our understanding of the state? The state carries out various activities and practices in its attempt at hegemonizing gender identities, practices which in turn have a bearing on the nature of gender identities among the subject population.[6] These activities and practices I divide into three broadly defined spheres. First, the state intervenes directly in the construction and consolidation of particular masculinities and femininities. Feminist literature on policies in welfare states shows how states reinforce a particular gender regime, through legislation and creating institutions for particular masculinities and femininities. In many Western states, these are centrally related to a femininity constituted around non-working housewife-mothers and a masculinity tied to the notion of wage-earners (Moore 1988: 128; Franzway *et al.* 1989; Pateman 1988; Fraser 1989).[7]

Second, the arena of state actions defines the boundaries between different masculinities and femininities, legislating on (un)acceptable masculinities and femininities, granting recognition and/or support to individuals who fall into appropriate gender roles, or adjudicating on acceptable gender identities. In other words, 'the state . . . has a role not just in regulating people's lives but in defining gender ideologies, conceptions of "femininity" and "masculinity", determining ideas about what sort of persons women and men should be' (Moore 1988: 129).

Lastly, state institutions incorporate only certain specific masculinities and femininities into its power apparatus as its agents. In this way, only the individuals corresponding to particular favoured gender identities gain access to or have an overt voice in the state. Although not using the terms masculinity and femininity, Carole Pateman (1988) has cogently argued that the Western concept of national citizenship has been constructed around a particular masculine model thereby excluding feminine roles (as well as alternative masculinities). Other writers, such as Joan Scott (1986), have highlighted how the state is deeply involved in discourses which use notions of masculinity and femininity to describe its own actions. On a more quotidian level, it has been shown how some states exclude certain groups of masculinities, such as gay and ethnic minority males, and many types of femininity, from normative images of power and authority figures.[8] Out of the three spheres of state action listed here, and which are crucial to the understanding of gender and the state, I focus on the first two, thereby attempting to provide a 'middle-level' analysis[9] (Alvarez 1990).

Conscious of their interests, women and men are active in the construction of their own identities, within particular constraints associated with the geographical and historical contexts of their action. These socially formed identities can be in resistance to state projects for shaping femininity or masculinity, as gender identity is not solely determined by the state (see Anthias and Yuval-Davis 1989: 11). Teresa de Lauretis suggests working with 'a conception of the [gendered] subject as multiple . . . as excessive or heteronomous vis-à-vis the state ideological apparati and the social-cultural technologies of gender' (de Lauretis 1987: x).

Before examining these issues in relation to a Peruvian case study, it is important to note that the relationship between the state and civil society is one of constant negotiation and contestation, between the state and social groups and between social groups themselves. The state attempts to impose certain masculinities and femininities, but its projects remain incomplete (see Mouffe 1979; Laclau and Mouffe 1985). Civil societal groups, whether based on cultural identification, class, religion or sexual orientation, resist the state's imposition and attempt to promote and consolidate their own gendered identities. The expression of non-state masculinities and femininities arises from civil society wherein facets of individual and group identity are expressed through gender (in practices or discourses). These gendered identities provide a basis for contesting state conceptualizations of gender regimes (Cocks 1989; de Lauretis 1987: 18).

STATE CONSTRUCTIONS OF FEMININITIES IN PERU, 1968–90

What type of femininities have been involved in the recent process of female peasant mobilization in Peru? What are the relations between state constructions of femininities and the response and resistance to this by peasant cultures? Given that peasant women have extended and consolidated various political identities in the past decade, what role has the Peruvian state had in this process? This chapter suggests answers to these questions.

State programmes for constituting gender identities are evidenced in a variety of means including legislation, symbolic use of gendered images, the type of organizations encouraged (e.g. charities, unions, the Catholic Church), government organizations on a local level, ministries established, education policies, and the ideologies of parties/groups in government. These dimensions of state action have varying degrees of impact on peasant women and men, the main focus of the chapter, although all are represented during the period under consideration. As Sonia Alvarez points out (1990), it is not only explicitly gender-related state programmes which have an impact on women, but *all* state activities.

The years 1968 to 1990 were characterized by the rapid growth of peasant confederations, concomitant with the increasing visibility of *campesinas* in the unions. As they rose in importance within the CCP and CNA, peasant women became engaged in setting out demands which at times challenged state-sponsored femininities, while on other occasions they selectively adopted state imageries in constituting their own gender identities. The three periods under consideration here are those under President General Juan Velasco (1969–75) in the first period, followed by Morales Bermudez (1975–80) and Fernando Belaúnde Terry (1980–5) in the second, and finally Alan García Perez (1985–90).[10] Each period encompasses different attempts by the state to construct particular femininities, and the diverse attempts at resistance and selective adoption by peasant women.

Incorporating peasants: the Velasco years 1968–75

The military government of General Velasco was unusual in Latin American terms in that it came to power with a radical programme of reform. Via its corporatist strategies and populist and pro-peasant rhetoric, it hoped to achieve a greater degree of national cohesion and security and to this end instituted a series of significant reforms in its early years. A major Agrarian Reform (1969) was followed by national-ization of many economic sectors (namely petrol, mining, agro-industry, fishing, banking, electricity, transport) and a series of state-led mobiliza-tion efforts in the countryside and shantytowns (notably via SINAMOS, or the National System of Social Mobilization) (Lowenthal 1975; McClintock and Lowenthal 1983; Slater 1989). In the countryside, the military government created a national confederation for the bene-ficiaries of the Agrarian Reform, called the Confederación Nacional Agraria (CNA: National Agrarian Confederation) in 1972. This con-federation provided support to the military government throughout the early years, although it subsequently split from military patronage in 1978 over the limited extent of the Reform and lack of support for the new agrarian cooperatives.

During this period, the state's definition of peasants envisaged them as male farmers with a particular class position. As a consequence of this, the state changed the term applied to Quechua[11] Andean peasants from the historical name *indígenas* (indigenous peoples) to *campesinos* (peasants). Class position was thereby declared by the state to be of greater importance than culture, a point readily accepted by the peasants whose experience of marginalization and cultural discrimination they wanted to end.

At the same time as attempting (generally unsuccessfully) to suppress the negative connotations of ethnic difference, the military government adopted two historical indigenous figures as major symbols in their rhetoric. These were Tupac Amaru and Micaela Bastídes,[12] an 'Indian' *mestizo* husband and wife who led a well-organized and bloody rebellion in 1781–3 against the colonial state.[13] The military government of 1968–75 was the first (and last) state in Peruvian post-Conquest history to adopt Quechua models of male and female partners in its presenta-tion of development plans. Government plans were called after the Inca or Tupac Amaru, and posters and images of the couple were widely distributed between 1968 and 1975. The symbol of SINAMOS, for example, was a stylized head of Tupac Amaru.

Such a model of state masculinities and femininities had important implications for peasant men and women's self-identification, as much due to its grounding within Andean peasant culture's conceptions of gender, as to its powerful ethnic-nationalist symbolism. By presenting the paired male-female group of Tupac Amaru and Micaela Bastídes,

the state (unwittingly) took on and elaborated a basic Quechua normative image of the *yanantin* or married couple. This notion of two equal
'mirrored' and gendered partners working together on dovetailed tasks
(for example, men ploughing, women planting seed; men in field tasks,
women in cooking) resonated with Quechua self-representation, while
at the same time elaborated and formalized this imagery in the forum
of the nation-state.

Regardless of specific (and limited) government imagery of peasant
women as equal participants in radical reform, in practice women were
treated as (yet another) homogeneous potential interest group to be
taken into the corporate state structure. Without questioning women's
gender-specific situation, the government proposed to 'incorporate
women into development' in a programme of promotion and equal
rights (Ruiz Bravo 1987; Bourque and Warren 1981). The United
Nations Decade for Women (1975–85) provided the international context for attention to be paid to women's incorporation into government
plans, although in Peru, few of the long-term planning proposals for
women were implemented. The Plan Inca of 1974, for example, called
for equality for women in rights and obligations, female promotion to
high posts, the elimination of discrimination, the promotion of coeducation, and common property rights for husbands and wives (also Comisión
de la Fuerza Armada 1977). However, none of these measures, save
coeducation in rural areas, was implemented (Bourque and Warren
1981: 184).

Nevertheless in a contradictory fashion, state apparati were also
subject to change through the actions of groups with alternative femininities. During the mid-1970s, the state officially recognized various
feminist study groups, which formed a Women's Front.[14] At the same
time, feminist lawyers rewrote the civil code in order to remove
discriminatory language (Bourque 1989: 128).

It was in other fields of legislation that peasant women's position was
most directly affected during the military government. The Agrarian
Reform legislation of 1969 had significant effects on peasant women's
economic position and status as citizens (Deere 1985). *Campesinas* were
effectively excluded from the Agrarian Reform programme, in both the
new cooperative structures and in the local administration of agricultural
enterprises. Within the new farms and cooperatives, women were not
granted beneficiary status and hence were excluded from land-holding.
Women in the beneficiary group numbered a mere 2 per cent, although
women continued to work in the fields and take on diverse agricultural
tasks for family or wages (ibid.; Bourque and Warren 1981: 184). In
village communities unaffected by land distribution, as well as in new
cooperatives, all adult women excepting widows and single mothers with
dependent children were excluded from administration on the basis that

they were already 'represented' by the male head of household. Single mothers who later married could also lose land to their husbands. The CNA confederation and its constituent Agrarian Leagues were likewise exclusionary of women, as they were based upon official membership of the rural enterprises created by the reform.

However, peasant women were also targeted specifically in order to create another corporate group within the state. An organization for peasant women the Asociación de Cooperación con la Mujer Campesina (ACOMUC: Association of Cooperation with Peasant Women), was created to mobilize *campesina* women and to channel their interests in parallel to the CNA.[15] Created in 1972, the same year as its coorganiza-tion, the objectives of ACOMUC were to 'promote the fundamental values of the peasant family through the education of family members, [and to encourage] all women to participate in society through the setting up of day-care centres [and] training programmes' (quoted in Deere 1977: 27).

As the above quotation makes clear, the programme's underlying expectation of peasant women was of domestic workers concerned with childcare with only limited 'participation in society'. Such a Hispanic urban femininity was (and is) alien to the *campesina* women who are active workers in the fields, market their produce in the local towns, and whose childcare responsibilities (only for those children under 6 years old, for after this age children start to work in childcare as well as around the house and livestock pasturing) are delegated to older children or are shared among female kin (Bourque and Warren 1981; Deere and León 1982). In other words, the state attempted to mobilize women on the basis of a femininity inappropriate to Quechua culture in the Andes. The implicit proposal of ACOMUC was reinforced by the initial staffing of the programme with middle-class urban women, often wives of Agrarian Reform officials, whose own femininity was more securely grounded in that articulated in ACOMUC's agenda. Nevertheless, by 1977, training programmes, in craft skills for around 3,000 *campesinas* had been started, especially in the cooperatives and surrounding peasant communities.

The disjuncture in ACOMUC between two femininities, that of Andean peasants and that of urban activists, created its own dynamic which undermined mobilization on behalf of the state and the state's project to create a different femininity in the countryside. Organizers and extension workers soon realized how articulate and politicized the *campesinas* were, as they demanded a part in community decision making and equal wages for their work. However, such points could not be addressed by ACOMUC as they fell under the remit of the Agrarian Reform legislation from which peasant women had been largely ex-cluded. For this reason, comments and criticisms did not go further than the organizers' reports, and this factor, combined with the lack of resources controlled by ACOMUC along with limited geographical and class extension (concentrated in cooperatives and among richer peasants), meant that it had no lasting impact on mobilizing peasant women. Nevertheless, the organization was significant for enlightening certain women of their position and encouraging them to participate.

Being non-unitary in its effects, different state representatives acted in contradictory ways to constitute gender identities. In some cases, national politicians designed and legislated for certain groups, although regional culture reshaped the meanings and practices of these groups. In other circumstances, local state officials blocked national plans they considered promoted 'inappropriate' femininities (see Afshar 1987). Several examples can be mentioned briefly. First, legislation from the Velasco government introduced army recruitment of women, an option which was picked by some *campesinas* whose self-representation did not exclude participation in the armed forces. However, when these women tried to register in the army, local state officials obstructed their applications as this threatened their own notions of appropriate femininity (Bourque and Warren 1981: 186–7). In a second situation, SINAMOS workers developed a femininity at odds with national-level rhetoric and legislation: female SINAMOS workers, local representatives of Velasco's corporatist state, attempted to consolidate feminist identities and practices which clashed with the passive rural female domestic worker perceived by Limeño legislatures. Finally, in other areas, peasant women attempted to consolidate their own independent groups despite the cross-cutting involvement of ACOMUC and state institutions such as CONAMUP (Dibos Cauri 1976).

The context for the constitution of *campesinas*' identities changed considerably with the bloodless *coup* of Morales Bermúdez, and then democracy.

'Mothering'[16] images: Morales Bermúdez and Belaúnde 1975–85

The period between 1975 and 1985 was one of increased economic crisis in Peru and of social mobilization at the grassroots to guarantee livelihoods. The state pulled out of reform and social mobilization at this time, preferring to let private assistance and the Church provide the arenas for development. Thus, although this ten-year period saw a military government followed by five years of democracy, state construction of femininities was similar throughout, as was the nature of resistance developed in response to it by peasant groups.

Where the state did intervene in constructing and consolidating a particular femininity was in emphasizing the mother–child dyad. The Comisión Nacional de la Mujer Peruana (CONAMUP: National Commission on Peruvian Women) was broken up during Morales Bermúdez's term to be replaced by the Instituto Nacional de Protección al Menor y la Familia (INAPROMEF: National Institute for the Protection of Minors and the Family) and the Comité de Revaloración del Menor (COPROME: Committee of Revaluation of Children) (Ruiz Bravo 1987: 99). Within this imagery 'The mother could not be a mother without being self-sacrificing. . . . The mother is the symbol of love, sacrifice and sanctity. . . . This [woman] is considered . . . as a religious symbol' (quoted in ibid.: 100).

The return to democracy in 1980 saw a continuing emphasis on low-income women as providers for their families and as mothers. Under the INAPROMEF, the state created infrastructure and distributed foodstuffs and gas stoves to shantytown women, organized in popular kitchens. INAPROMEF (later INABIF) functioned under the auspices of, and often with the active participation of, the President's wife whose European model of femininity was closely associated with charitable benevolence and modesty.

Assistance programmes in shantytowns were concerned to educate and train women in nutrition, infant psychology, and budgeting for food purchases (Barrig 1986: 168). Although Belaúnde's government had no explicit policy concerning women as women (Ruiz Bravo 1987: 114; Portocarrero 1990: 113), the state process of consolidating particular femininities continued within the broader political culture.[17] Treating low-income women as mothers became the prevalent discourse and practice. As one commentator observes, 'Maternity [was] the most important role taken by women, including its economic aspects' (Barrig 1986: 172). The government consolidated this pattern of femininity by removing support from the Mothers' Clubs (groups of women meeting to carry out craft or agricultural work) and Church groups which they considered to be 'too political', that is those who demanded to be treated as active agents in their own right. This selective cooptation of low-income women's groups, according to their confirmation of state-sanctioned femininities, continued with the election and inauguration of García's social democratic government in 1985.

Alan García and the continuation of mothering roles, 1985–90

State attempts to consolidate a 'mothering' role for low-income women continued under the government of Alan García and the Alianza Popular Revolucionaria Americana (APRA: American Popular Revolutionary Alliance). However, differences in this party's history translated into distinct strategies in the constitution of low-income mothering femininity. APRA's ideology and history of patronage was founded upon a highly-gendered imagery. Modelled upon a hierarchical family structure, APRA's strength was historically founded upon entire nuclear families joining the party as activists (Barrig 1986: 153). In this 'family discourse', traditional and internal hierarchies of the patriarchal family are reinforced and subsequently reinscribed in government ministries and party divisions of labour. For example, women were given ministerial posts for the first time in Peruvian history under García, but were limited to the fields of health, education, and so on. However, this party history perhaps explains the larger number of female officials in this party than in other political groupings in Peru.

The broadly social democratic APRA government extended and consolidated its own femininity in relation to rural women through assistance and work-creation programmes, emphasizing once again the importance of mothering in its dealings with low-income women. Distribution of milk powder to mothers with young children in the Glass of Milk programme, and the creation of Mothers' Clubs for local communal work programmes were two major aspects of APRA practice. Through these assistance programmes, peasant women as well as shantytown dwellers were regarded in a passive mothering role. Mother and child programmes were presented by government spokeswomen as one of the greatest government-initiated benefits for women (*Mujer y Sociedad* 1987: 6). While shantytown women in Lima and elsewhere disputed these attempts by APRA to incorporate their means of livelihood into state agendas by means of demonstrations and independent organization (Radcliffe 1988), the economic situation was such that for survival reasons they often had to assist in these organizations (Graham 1991). Similarly, in work-creation programmes such as PAIT,[18] shantytown women did not accept the construction of low-income femininity integral to their implementation. In the countryside, peasant women were similarly subject to cooptation into party-related organizations, such as work programmes and Mothers' Clubs.

Nevertheless, as in previous periods, the state's impact on gendered identities was complex and often contradictory. At the same time as consolidating a 'mothering' low-income femininity, the state increasingly recognized women's labour market participation. García oversaw legislation on a variety of middle-class concerns, such as sexual harassment at work in 1986, and a new Civil Code in 1984 which emphasized the rights of the entire family rather than just those of male adults (Ruiz Bravo 1987: 118–9). However, more radical ideas proposed by the few women senators, such as social security cover for women working at home and loans for women, were not passed (*Mujer y Sociedad* 1987: 7), and a liberal abortion law was drafted and then rejected.

The southern department of Puno serves as an example of the relationship between peasants and the state around which gender identities were forged during these years. First, the Programa de

208

Asistencia Directa (PAD: Direct Assistance Programme), a pet project of the President's wife and run from the Presidential Palace (Graham 1991), was based on three principles: a respect for the institution of the family; saving of mothers' time, and the avoidance of infrastructure construction. While not explicitly targeted at women, the PAD recognized women's reproductive work and the importance of women's income to family welfare, and worked through women's organizations such as Mothers' Clubs, craft workshope and nurseries (Portocarrero 1990: 115–16). Thus, the PAD programme reached *campesina* women in Puno. Simultaneously, APRA placed party members in charge of the clubs, a tactic recognized by the *campesinas* in the region who were unable to find alternative funding for their groups, due to the withdrawal of food aid by the private charity CARITAS (interview L.S.[19] 1990). As sources of support shifted, women directed their claims to APRA-party state officials: delegations of peasant women made demands at the town hall in Puno in February 1990 for the Glass of Milk programme to be extended to their villages.

Another factor was the rapid rise in living costs resulting from high inflation, which in 1990 reached over 7,000 per cent annually, and the imposition of emergency economic measures which removed food subsidies on goods (e.g. the *paquetazo* of September 1988), in desperate attempts to revive the collapsing economy, which contracted by 11 per cent in 1989 alone. Living standards throughout the country plummeted: in 1989, it was estimated that the poorest 40 per cent of population received only 6.8 per cent of the national income[20] (*Situación latinoamericana* 1991: 159). Faced with worsening poverty for their families, peasant women in Puno were active in representing their interests to local and national government. In September 1989, a 'sacrifice march' took place, involving 8,000 people of whom 60 per cent were women and leading to the occupation of the town hall in the department capital (interviews J.R., L.S. 1990). Among demands for better prices for their products, adequate health and educational provision, *campesinas* were also vociferous in asking for equal rights for women (interview S.P. 1990).

Throughout the García regime, peasant women and men were involved in reclaiming agricultural land from the state-run cooperatives, by means of land invasions. Viewed as serious threats to order, the state retaliated with force and *campesinas* faced tear-gas and arrest for defending the land reclaimed. Up to 3,000 peasants took part in land invasions in the Benavista farm in 1989 and 1990, resulting in the death of one *campesina*, broken hands for one man and one *campesina*, and hospitalization for tear-gas effects of Damiana Juna Quispe of Coata (interview FDCP 1990).

PEASANT RESISTANCES AND GENDER IDENTITIES[21]

How did peasant women understand, react to and negotiate their identities in relation to the above practices undertaken by the state during the 1968–90 period? Generally the state, largely unwittingly, contributed to a reassessment by peasant women of their place both in relation to the state institutions and imagery to which they had access, but also in relation to their own peasant organizations.[22] Moreover, peasant women's relationship to the state practices of gender constitution were often distinct to that of peasant men, due to their distinct agendas and priorities in identity construction.[23]

One example of this is the pattern of re-presentation of the indigenous couple, Tupac Amaru and Micaela Bastídes, by peasants. The falsely homogeneous imagery of 'indigenous' masculinity and femininity was elaborated by *campesino* women and men after its deployment by the military government (which withdrew from its *indigenista* rhetoric after 1975). With growing pressure from the state on women to conform to a housewife–provider role from the early 1970s, *campesina* women resisted in their actions, citing Micaela Bastídes (and to a lesser extent Tupac Amaru) in a warrior role as examples and forerunners of their contemporary struggle. As one female peasant leader said, 'People have to rise up – like the great women fighters . . . Micaela Bastídes, Clorinda Matto de Turner[24] . . . who lost their lives' (interview A.Q. 1988). By contrast, male peasants tended not to cite these figures in their description and analysis of the situation of the peasantries.

Another difference is that women, utilizing the historical figure of Micaela Bastídes as a template or ideal, negotiated their relationships within the union by taking her as a precedent for women acting as political partners in a male–female couple. One *campesina* explained, 'What courage these women [Micaela Bastídes, women in the conquest] had! Why not fight for another future, to be equal to others? Men and women fight together, united, sharing . . . it can be better. Micaela Bastídes surely fought for her children' (interview A.Q. 1988).

With rising educational levels in the communities, peasant women have become increasingly aware of other struggles around the country and in Peruvian history. Moreover, learning Spanish allowed them to understand the radio and the media. Nationally, rates of female rural illiteracy fell from 68 per cent in 1972 to 53 per cent in 1981 (Cavassa and Portugal 1985: 71). Among the female peasant leaders interviewed (and for whom educational experience is recorded), levels of schooling are relatively high at four and a half years on average. However, nine women born since the 1940s have five years education on average, compared with the two women born before this date who received no education as children. This is reflected in an increasing engagement by peasant women with state symbols and histories.

Peasant adoption of the imagery of Micaela Bastídes and Tupac Amaru highlights the lack of control by the state over cultural forms, as well as the possibility of resistance through them to state projects (see Rowe and Schelling 1991). The utilization by *campesinas* of Micaela Bastídes' example further illustrates the (gendered) ways in which cultural forms are constantly misrepresented and reinvented.

State institutions also contributed (unintentionally) to the renegotiation of peasant women's self-perception. For example, ACOMUC played a role in awakening peasant women to alternative femininities in relation to the state. One *campesina* explained her understanding of its role in the following terms: 'The ACOMUC, at the time of the Agrarian Reform, worked with us with leaflets. It was very important – showing us how to value ourselves as women. There hadn't [previously] been valorizing of women's participation' (interview E.N. 1990).

Despite peasant women's increasing awareness of precedents for activism in politics, participation in the confederations in the late 1970s and early 1980s involved women in helping male peasants in subordinate roles. During the 1970s, peasant women and men mobilized to protect peasant economic interests, within the context of a peasant-specific and non-state gender regime. This was a gender regime of male–female 'partnership', characterized by male public speaking and decision making, and female participation in supportive roles. Such gender identities assumed that male leaders presented plans and played the major public role in the unions, while in the domestic sphere women and men were equal partners. Official union discourse about gender relations expressed the belief that men were able to speak on women's behalf, since the two genders shared the same discrimination and disadvantage *vis-à-vis* the state.

On this basis, a Secretariat of Women's Affairs was created in the CCP in 1979 and the post held by a man for three years; the same happened in the CNA a few years later. Reinforcing the legitimacy of this move was widespread pride in rural popular culture and a rejection of urban Hispanic modalities of life. Most importantly, the hierarchical gender regime did not prevent the continued mobilization of peasant women in larger and larger numbers in the confederations. In these early years of female involvement, women were found disproportionately in the sections of the union concerned with 'female issues', and held posts such as Secretary of Women's Organization, or Secretary of Women's Affairs in district and departmental unions.

At the same time that state rhetoric (of Bastídes and Amaru) entered the discourse of *campesinas*, the peasantry was manipulating the gendered imagery and practice embedded within state institutions. In other words, emphasis on class and ethnic roots to the conflict with the state did not blind peasants to the distinction between their own femininities

211

(and masculinities) and those practised in state institutions. This resulted in a dual pattern: on the one hand, peasants acted within a framework defined by a (Hispanic, urban, Catholic) image of women as weak dependants in their dealings with state agents such as the police, while on the other hand they consolidated a distinct gender regime within the unions, as noted above.

During the late 1970s, peasant women and men were active in reclaiming agricultural land which had been allocated to appropriated farms and cooperatives, and which often had originally belonged to the surrounding peasant communities. These land invasions (*tomas de tierras*) were crucial in the politicization process of peasants generally, and of *campesinas* in particular (García-Sayan 1982; Radcliffe forthcoming). These *tomas* were also occasions when distinct concepts of femininities were articulated and utilized. During the land invasions of 1978 and 1979 in Cuzco department, for example, peasants exploited a more Hispanic conception of femininities (and masculinities) prevalent in the police and army. Soldiers and police facing the peasants in land conflicts acted upon a masculinity which ostensibly forbids public violence against women including ethnic minority women. Women, expected to be more passive and weaker than men, were thus unsuitable targets for violent treatment from soldiers. The peasants had anticipated this response and, as they hoped, the police were unwilling to shoot at peasant women who formed barricades around *campesino* men, while the latter reclaimed the land and ploughed it (various interviews 1988; interviews 1990; also García-Sayan 1982; Radcliffe forthcoming). Peasant men, by contrast, were considered to be fully socially responsible and legitimate targets; they were shot, injured and arrested, while the *campesina* women were simply arrested.

Throughout the early 1980s, there was a notable increase in formal participation of women in union activity and leadership. With changing state policies towards peasants and women, *campesinas* were also transformed into active participants in the informal labour force, and a great variety of political fields. With the collapse of state provision of support for low-income families, women took on an increased burden of income generating, reproductive, health and welfare, and managerial work.

The situation of marginalization of peasants continued after 1985. Under the APRA government, only tentative efforts were made to provide the peasantry with the means to develop the agrarian economy, and the confederations continued to oppose government policies. Moreover, within the peasant confederations, the impact of the Aprista-inspired construction of femininity had a minimal effect, due both to peasants' consolidation of their resistance to the state and to the influence of left-wing opposition parties. These parties, incorporating issues propounded by urban feminists and by a popular low-income femininity developing in the shantytowns, provided a space for a more combative, openly politically committed femininity. It was this femininity, which I call radical *campesina* femininity, which provided a political space for *campesina* leaders during the 1980s, and allowed them an identity as political actors, one dimension of which was a rejection of APRA-inspired clientalist positions, as discussed below.[25]

In relation to the confederations' own gender regimes, peasant femininities remained the same, although women comprised a larger share of union membership. Generally during the mid- to late 1980s, the earlier gender regime of male–female partnership continued. Nevertheless, the profile of gender-specific issues moved higher during the late 1980s, due to an increasing recognition within the unions that women played a crucial role in family survival under conditions of economic crisis. The confederations began to adopt policies to equalize membership among male and female peasants, and to structure meetings to allow both sexes' participation (by providing separate rooms at conferences, and having women cook).

Such changes resulted in high profiles for women in confederation structures. In 1987, the CCP created a National Committee of Women's Affairs with eight regional representatives, in order to coordinate the mobilization of consciousness raising of peasant women throughout the confederation structure.[26] National and regional meetings of grassroots delegates took place: the First National Meeting of Peasant Women took place in Lima in April 1987 with 400 representatives, and subsequently numerous local meetings occurred throughout Peru. By 1989, women were found in the highest executive councils of the CCP and CNA confederations: the CCP had two women in the National Executive Committee of twenty-six members, while the CNA appointed two women in senior positions in its more flexible structure. The majority of female leaders continued to stress the non-gendered aspects of their identities as peasants, inheritors of land and members of the popular classes, rather than feeling themselves subsumed within a straight-forward identification as 'women' (Radcliffe 1990a). Nevertheless, rural women in Puno voiced gender-specific critiques of the state and its effects, charging political parties with using them in election campaigns and then forgetting them, by treating them just as housewives despite their varied income generating occupations, and for failure to grant official recognition to women's organizations (FDCP n.d.).

Feminism, although relatively influential in urban areas, has generally been rejected by peasant women as it is perceived as a subject position available only to urban and wealthier women. One peasant woman said,

It isn't like the feminists say. . . . Women have to be prepared to, and know how to, understand their husbands, have patience. . . .

In the cities, men and women separate, women paint their nails, their lips, go for tea. I don't agree; I'd rather be poorly paid. I'm shocked by the feminists and I tell them so.

(interview C.Q. 1988)

Nevertheless, a handful of female leaders challenged what they described as their minor, supportive role within the unions, and articulated a radical *campesina* femininity. By citing Micaela Bastídes as a legitimating example for their actions, these *campesina* women began to criticize confederation structures and policies for ignoring many of women's demands and concerns (Radcliffe forthcoming). Certain women left the confederations to work more closely with feminist organizations, such as the National Rural Women's Network run by the Lima-based feminist group, Flora Tristán (interview B.F. 1988), or independent non-governmental organizations (interviews A.Q., J.R.P. 1990).

The rise of a current within the peasant female leadership which questioned the roles taken by men was not unique. Other, distinct femininities were present in the unions by the early 1990s, each negotiating a place for itself within confederation structure in relation to the state and to everyday practice in the villages (Radcliffe 1990a, forthcoming). As noted above, the state provided an arena for contradictory influences on gendered identities, despite its attempts at providing a coherent model for various groups. In this way, the influence of the state varied, according to the differential spatial effects and the hegemony of its projects. Among the influences mitigating the femininities generated by the central state were regional variations in the state's institutions and effects, alternative ideologies (such as Liberation Theology, feminism, and so on) and regional cultural differences. Due to these factors, the uneven spatial distribution of femininities within Peru arose. *Campesina* femininities in Puno department, for example, were distinct from those in other regions due to the spatially uneven impact of the nation-state and to the cultural specifics of the department, which gave rise to particular forms of resistance in peasant femininities.

In Puno department, a group of *campesinas* created an Asociación Departamental de Mujeres Campesinas (Departmental Association of Peasant Women) affiliated to the local CCP-linked Federación Departamental de Campesinos de Puno (FDCP: Departmental Federation of Peasants of Puno). The creation of the Association in 1985 from the pre-existing Interdistrict Committee of Peasant Women (founded 1981) highlights the *campesinas'* wish to maintain a women-only organization, and distinguishes them from other regions' women who joined mixed federations.

Not only did Puneña *campesinas* retain semi-independence, but they also intended to create their own federation completely separate from

214

the FDCP, causing considerable tension between the Association and the departmental FDCP (interviews L.S. 1990). At the First Departmental Congress of Peasant Women in May 1985 attended by 500 *campesinas*, the motion to create a women's federation was blocked by the main left-wing party within the FDCP (interview J.R.P. 1990; see PUM Comisión Femenina 1985). By the time of the Second Congress in June 1990, the peasant women maintained their priority of an independent group, although with the declining economic situation and increasing rural violence the importance of coordination with the FDCP appeared to be winning over. Moreover, women from the Association continued to hold posts in the departmental federation. For example, the leader of the Associación Departamental de Mujeres Campesinas also held the posts of Secretary of Women's Affairs and of Secretary of Education and Organization in the FDCP.

The dimensions of this regional femininity extended beyond a concern for an independent women's group, and included religious, cultural and regional facets. The women comprising this group identified themselves as Christian women in the specific regional culture of the Southern Andes. Articulating a conception of themselves as women (*mujeres, warmi*), Puneña female leaders created a femininity in which cultural and social discourses and practices were distinct to those represented elsewhere in Peru.

Compared with the less localized, highland identities articulated by other female union leaders (Radcliffe 1990a), Puneña women identified themselves specifically and locally as Quechua or Aymara. By retaining their languages and clothes, they explicitly projected strong ethnic markers compared to other women in the confederations. One woman explained how they always wore '*ojotitas y nuestros vestidos típicos*' ('rubber sandals and our typical clothes') (interview L.S. 1990). Moreover, this same woman learnt Aymara as well as Spanish as an adult, in order to work with Aymara-speaking communities in the south of the department. The reasons for this cultural pride are not straightforward, but would appear to be related to the historical construction of the Southern Andes as the geographical centre of Otherness in Peru by successive governments and intellectuals (such as the Indigenists movement of the 1920s). Another factor is the part played by the dynamic ethnic rights groups in neighbouring Bolivia epitomized by the Katarista Peasant Movement (see, for example, Rivera Cusicanqui 1987; Calderón and Dandler 1986). The current leader of the *campesinas*' association met the Bolivian miner women's leader Domitila Barrios de Chungara in 1985, and maintained an interest in low-income women's groups in Bolivia. Considering the Puno women then, the ethnic distinctiveness of their regional grouping was consolidated by the merger of Quechua and Aymara women's groups in Puno. Although the original group

215

comprised Quechua-speaking women, in 1990 they began to coordinate with the smaller Aymara women's group, the Comité de Mujeres Aymaras '8 de marzo' (Aymara Women's Committee '8th of March'),[27] founded 1981.

A second element in the Puno *campesinas*' femininity, their Christianity, is easier to explain. The Liberation Theology Church and its associated Christian base communities were active in Puno from the 1970s onwards (CEAS 1985), organizing peasant and low-income urban women into self-support groups. Church-led discussions around women and their role in the community highlighted such issues as women's right to demand equal respect; men and women's equal value before God;[28] and how to conduct meetings and carry out community business (writing official letters, filling out receipts, taking notes and maintaining order at a meeting) (Instituto de Estudios Aymaras 1985). Women were also taught about women acting as leaders in the Bible (Deborah) and Peruvian history, and were introduced to the experience of contemporary women leaders such as Domitila Barrios de Chungara. For example, using quotations from the latter's history (Barrios 1978), women were told the 'true story' of how Domitila taught her husband the equal value of housework and her union work (Instituto de Estudios Aymaras 1985).

With their confidence raised by such examples and a grounding in organization, Puneña peasant women became involved in federation work. For these women, religious work often predated union activity. One woman had a long history of activism within the local Church networks from the age of 15 years, taking up official union posts only at the age of 38 years. The Association *campesina* leaders recognized that their work on behalf of the union was similar to their evangelical counterparts' actions. However, they also recognized and deplored the limited extent to which local parishes were willing to fund women-only groups at the cost of community projects (interview S.P. 1990).

Third, these *campesinas* identified strongly as women, not in a Western feminist sense which they rejected, but in terms of wanting independent *campesina* organizations. As noted above, in this latter respect the Puneña women broke with existing national confederation policy and the majority of female leaders, who argued for the integration of men and women in peasant unions. Their prioritization of an independent group may have been influenced by the existence in Bolivia of a *campesina* women's federation. The Bolivian Federación de Mujeres Campesinas 'Bartolina Sisa' (Peasant Women's Federation 'Bartolina Sisa') provided a vivid contrast to the Peruvian situation (Sostres and Ardaya 1984; Muñoz 1986), while visits were arranged by Peruvian *campesinas* to the Bolivian federation.

In their discourse of radical *campesina* femininity, women's relation to politics was represented as fundamental to their beings, from their birth; they could not help but be political. One Puneña argued,

216

Yes, we women have been born with politics. When we're born, immediately we shout, we demand things, don't we? For this reason, we women have been born with politics. So what is it to talk about politics? It isn't difficult; the dog always barks.

(interview L.S. 1990)

In this way, women claim that men and women have equal rights to a political voice and a role within peasant organization. The same woman explained,

Perhaps the Spanish brought over the marginalization of women. Some [peasant] men are like the Spaniards, ordering women 'Cook this, do that!', and so on. But when there's organization and consciousness, both [men and women] are cooking, helping get education for the children.

(interview L.S. 1990)

Despite their activism in gender terms, Puneña women remained strongly populist (indeed 'peasantist'), highlighting the need for realistic prices for peasant produce, social security and credit from the state. It is for this reason that the Association's leader was chosen as a candidate for Deputy in the national elections in 1990 in the list of the Partido Unificado Mariáteguista (PUM: United Mariateguist Party), part of the United Left coalition. She was not elected, not least because of the lack of attention given her by the mainstream media, concerned about her lack of fluent Spanish and her *campesina* image (interviews L.S., S.P. 1990).

CONCLUSIONS

Joan Scott (1986: 1067) has argued that 'politics constructs gender and gender constructs politics'. This chapter has attempted to illustrate these points by examining state constructions and re-presentations of femininities, and the resistance to these by groups in civil society. I have argued that state-sanctioned constructions of femininities feed into political culture, in turn impacting on the nature of political activism in peasant Peru in a constant flux of negotiations, appropriations and negations. Between state-sanctioned masculinities and femininities on the one hand, and 'popular culture' urban and low-income femininities and masculinities[29] on the other, peasant women have constituted their gendered identities and articulated their interests.

Over the period under consideration in this chapter, the Peruvian state attempted to hegemonize, via its programmes and legislation various ideological femininities which were generally remote frr peasant women's concerns and identities. In response, peasant wr

217

expressed gender interests in culturally- and class-specific ways, in ways which reinvented and rearticulated state-sanctioned femininities, and, in cases, in ways which resisted them. However, the process through which they forged their femininities was in dialogue with the state. The Velasco government's successful interpolation of the figurehead of Micaela Bastídes with peasant women had the most long-term impact on peasant femininities, although the way in which this element of political culture was elaborated was unforeseen by state and peasantry alike. By contrast, the attempts by later governments to institute an alternative, more urban Hispanic identity were rejected by *campesina* leaders.

Overall, the state is an important arena in which the construction of femininities and masculinities takes place in the political culture of Peru. However, it is also an actor in the constitution of gendered identities, although its actions (because carried out by differently placed individuals) are at times contradictory and their consequences unforeseen. Protest arises then not only in response to the state's economic and 'formal' political agendas, but also in the sphere of political culture.

NOTES

2 THE SEEKING OF TRUTH AND THE GENDERING OF CONSCIOUSNESS

1 Interviews were conducted by the author in Spanish with the CoMadres in San Salvador, Caracas and Washington between 1987 and 1991; interviews with CONAVIGUA were conducted in Guatemala in 1991.
2 This interview with María Teresa Tula was conducted in Washington by Gretchen Bosschart whilst a Ford Honors student under my tutelage in Women's Studies at Wellesley College in spring 1988.
3 FMLN stands for the guerrilla coalition movement, the Farabundo Martí National Liberation Front.
4 América Sosa was interviewed by the author in San Salvador in November 1987 at the Congress of FEDEFAM (Federation of Relatives of the Detained-Disappeared in Latin America).
5 Marianella García Villas, President of the Non-governmental Human Rights Commission and a lawyer who had been actively gathering testimonies and evidence of human rights violations by governmental security forces, was assassinated on 14 March 1983 by the First Infantry Brigade during a trip into the countryside to gather evidence on the military's use of phosphorous bombs and biological germ warfare against the Salvadoran population.
   CODEFAM 'Marianella García Villas', or the Committee of Relatives Pro-Freedom for Prisoners and the Politically Disappeared of El Salvador 'Marianella Garcia Villas', was formed in her honour, and now has more than 400 members. Together with COMAFAC, or the Committee of Christian Mothers and Relatives of Prisoners, the Disappeared and Assassinated, 'Father Octavio Ortiz and Hermana Silvia' and the CoMadres, an umbrella organization called FECMAFAM 'Monseñor Oscar Arnulfo Romero' (Federation of Committees of Mothers and Relatives of the Imprisoned-Disappeared and Politically Assassinated of El Salvador), has been formed to coordinate activities of all three groups.
6 The ARENA party represents the extreme right-wing interests of the ruling fourteen families in El Salvador. Under the ARENA-backed Cristiani government, the Anti-Terrorist Law was passed which allows for a fifteen-day detention of anyone suspected of 'terrorist' activities.
   For further reading on El Salvador, see McClintock 1985a and Armstrong and Shenk 1982.
7 I would like to thank JoAnn Eccher for providing me with some of the CONAVIGUA materials.
8 For more information on the scorched-earth strategy of the Guatemalan

y during the early 1980s, see Black *et al.* 1984; also, regarding the
ng of the Guatemalan military elite, see Schirmer 1991.
umes of CONAVIGUA widows are protective pseudonyms.
view by author in Guatemala, July 1991.

11 The office of the Guatemalan Human Rights Ombudsman estimates that the political repression in the early 1980s left 100,000 dead, 1 million displaced, 45,000 widows, 45,000 orphans, and 45,000 refugees (primarily in refugee camps in Mexico) (interview by author, 1991).

12 The UASP is a coalition of social, human rights and labour organizations formed during the early years of the Cerezo government (1986–90) to make demands on that government for reforms. For reading on popular resistance organizations and state repression in Guatemala since the 1960s, see McClintock 1985b and Fried *et al.* 1983.

13 Civil patrols, originally organized in the late 1970s by the guerrilla in the highlands to protect the indigenous population from the army, was expropriated by the army in 1982. Today, there are over 900,000 mostly indigenous peasant men who are forced to patrol their village, stopping people going in and out, and acting as the eyes and ears of the army, and are, in effect, an extension of the army surveillance apparatus. Six thousand of those who have refused to patrol have organized themselves into Consejo de Comunidades Etnicas Runujel Junam por el Derecho de los Marginados y Oprimidos (CERJ: Council of Ethnic Communities 'We Are All Equal' for the Rights of the Marginalized and Oppressed), with many members threatened and over twenty-five members assassinated or disappeared since July 1988 (CERJ 1991). (For further reading on the civil patrols, see Americas Watch 1986 and 1989.)

The other kind of forced recruitment is into the army itself, whose strategy is to forcibly pick up young men in villages and marketplaces and induct them into the army. (See flyer protesting such forced recruitment from the demonstration by CONAVIGUA widows and CERJ members in Santa Cruz del Quiche in June 1991.)

14 Guatemala has, next to Haiti, the second highest infant mortality rate at 79 per 1,000, and the lowest immunization (and tax) rate in all of Latin America; 66 per cent of the population have no access to health care of any kind.

15 For more information, see Amnesty International 1991 and Americas Watch 1991.

16 Military commissioners are civilians hired and armed by the military to gather intelligence and assist in recruitment. These two men in particular had been threatening the family because of its membership in CONAVIGUA and CERJ (see Americas Watch 1991).

17 Maxine Molyneux has provided feminists with a vocabulary for women's different forms of organizing as sources of strengths. She distinguishes between 'strategic gender interests' and 'practical gender interests'. The first obtains to strategic objectives by 'feminists' to overcome women's subordination; the second arises from concrete conditions of 'women' (not feminists) and represents a more 'tactical' 'female' need (Molyneux 1985).

18 While individual relatives of various groups of relatives (such as the President of GAM in Guatemala) have gone on to study law to be able to enter the legal and primarily male world of human rights, most of the relatives of the disappeared have not made the connection of demanding 'rights' as either citizens or women. They have been forced to learn the legal language

through which their governments address social issues, and thus it is not surprising that the language of life and death within a national security regime – without any reference to 'citizenship' – dominates the Relative-vs-Government debates. This legal emphasis conflicts with the collective consciousness of these groups in which social cohesion and grief-support networks tend to override any focus upon individual rights, and yet the issue of women's collective rights is also ignored.

19 In other interviews with the Plaza de Mayo Mothers of Argentina, the major spokesperson, Renee Epelbaum, told me in regard to the internal uprising of the military in 1989, the Tablada 20, 'We need to understand why this occurred without feeling that we need to either condemn it or justify it.' How mothers' groups in Argentina, Chile, Guatemala and El Salvador persist in asking 'why?' during democratic regimes and transitions is the subject of a forthcoming book by the author.

20 FEDEFAM, the Federation of the Relatives of the Detained-Disappeared, is constituted by committees in twenty-one countries. It holds an annual congress in one of these countries each year, with its headquarters in Caracas. For other examples of this gendered consciousness among the Plaza de Mayo Madres, the GAM, and the Chilean Agrupación see Schirmer 1988.

21 *Dilemma of the Feminist Ethnographer: Whose Voice? Whose Experience?* Having chosen not to assume the female/feminist and pragmatic/strategic dualisms, I am still faced with a dilemma: can I, as feminist ethnographer, 'set aside' my own assumptions, and let the women speak for themselves to see how they see their consciousness forming? To understand what they see are the immediate causes for this consciousness? To ask why they are able – and not others with the same experience – to make the connections between rape and the state? To see how they view 'feminism' to be constituted? Can my own analysis add to an understanding of the whys and hows of this particular gendered consciousness, without 'sanctifying the resulting melange' of women's actions as self-evident (Riley 1991), or without homogenizing and essentializing women's identity? Can ethnographers/anthropologists 'construct subjects' and 'historicize experience', as Joan Scott asks, without 'assuming that the appearance of a new identity is not inevitable or determined, not something that was always there simply waiting to be expressed, not something that will always exist in the form it was given in a particular political movement or at a particular historical moment'? (Scott 1991: 792, 797).

3 *ECOLOGIA*: WOMEN, ENVIRONMENT AND POLITICS IN VENEZUELA

1 The information on which this chapter is based comes from the experiences of the author in the multiple roles of activist, observer and analyst of the social movements during the last fifteen years, particularly the environmental and the feminist movements. During this period, the author organized several national and international seminars and meetings for the study and evaluation of the political perspectives of social movements in Venezuela. Direct sources of information were: participant observation, semi-structured in-depth interviews, self-evaluation sessions where each actor and organization developed its own diagnosis according to shared parameters, and negotiation workshops of evaluation. We used the methodologies of strategic planning and 'investigation' which we believe allow for the participation in the diagnosis and evaluation of the social actors involved.

221

2 As a consequence of the economic crisis and the acute deterioration of life conditions in the squatter settlements, men have adopted some problems traditionally associated with women's role as theirs. This has been the case in the latest demonstrations in Caracas, Barinitas and Guarenas against the severe water shortages. See the newspaper *El Nacional*, Caracas, 6–24 January 1992.

3 Throughout this chapter, the terms 'environment' and 'ecological' will be used as synonymous.

4 In low-income neighbourhood associations, it is difficult to establish the difference between the primary and the secondary demands because they tend to coincide.

5 Interview with selected women leaders of ecological and urban organizations (USB-ILDIS 1987).

6 Different from the middle-class neighbourhood associations, the low-income ones were born already politicized or under the tutelage of the political parties. The Nueva Ley Orgánica del Regimen Municipal (MLORM) which was approved in 1989 only exacerbated these trends (García Guadilla 1990).

7 'New political facts' are defined as those that are constructed in the terrain outside the conventional political parties and are allied with the new social movements.

8 An example of this is the mobilization for the discussion and change of the Código Civil. Maybe this was because the changes affected men and women alike.

9 The clientelism prevalent in the political system along with the electoral system established by the 1961 Constitution, which privileged voting by 'a party' and not by individuals, contributed to the strengthening of the parties. These parties agreed upon socio-political pacts in order to share power amongst themselves and the rest of the social forces, thus avoiding conflicts that could destabilize the political system.

10 The mistrust and rejection of political parties has entailed a questioning of the near-monopoly on power that parties retain but not of the basic rules of democracy or the political regime. In fact one of COPRE's proposals for reform was the internal democratization of the political parties. Despite the parties' initial resistance to self-democratization, the high rate of electoral abstentionism and null votes (reaching 70 per cent) in the regional and local elections of 1989 forced them to accept some of the proposed reforms.

11 President Lusinchi's speech at the founding ceremony of COPRE, Caracas 1984.

12 This information is based on the self-diagnosis made by the Grupo Miércoles (1989).

13 CONG 1988. Official document (free translation).

14 This information is based on in-depth interviews with feminists 1978–9 and self-diagnosis from *Las Primeras Jorenadas de Evaluación de los Movimientos Sociales en Venezuela*. (USB-ILDIS) 1986. The ideological origin from the political left and/or from the progressive intelligentsia is a characteristic shared by Mexican and Brazilian environmental groups (Viola 1987; Slater 1985; Quadri 1990).

15 Venezuela was perhaps the first country in Latin America to institutionalize, since the 1970s, environmental issues through legislation. In 1976 the Ley Orgánica del Ambiente (LOA: Organic Law on the Environment) and the Ministerio del Ambiente y de los Recursos Naturales Renovables (MARNR: Ministry of Environment) was created to watch over

compliance with the LOA. Within the legislative power, the Comisión de Ambiente y de Ordenación del Territorio was established with representatives from various political parties. At present, pending the final signature of the President, the Ley Penal del Ambiente will reinforce the above laws.

16 This apparent contradiction is due to the lost radicalism of MAS and its increasing transformation into a liberal party that no longer questions the government from the structural point of view.

17 Definitions of primary versus secondary social relationships are taken from Parsons 1951.

18 Rally Transamazónico. In April 1987, it became known that the organizers of the Paris–Dakar Automobile Rally were busy promoting a Trans-Amazonia Rally, which would start in Venezuela, cross the Amazon, and finish in Brazil. Immediately, AMIGRANSA mobilized and gathered extensive information about the situation, and denounced to the authorities the negative environmental impact that the rally would have upon one of the most fragile ecological regions of the world. Simultaneously, they secured the support and solidarity of national and international environmental organizations, and launched a far-reaching campaign which was to enjoy wide acceptance in the mass media. As a result of these actions, the government suspended the permit already granted for the rally. In turn, the rally organizers opted for changing their strategy. This time they requested permission for a Trans-Andean rally, still to start in Venezuela. Once again, the government rejected the new proposals, based on the environmentalists' arguments against the rally, and due to the strong pressure exerted by the environmentalists via print media.

About the Spilberg-Tepuyes issue and its evaluation by AMIGRANSA, see Boletín de los Grupos Ambientalistas No. 4. 1991.

19 Interview with Miri Liffschysta, member of CONG. Caracas, 13 November 1991.

20 Quinto Encuentro Latinoamericano de Mujeres. Buenos Aires, Argentina 1990. For the main topics discussed in these Encuentros see Encuentros Feministas Nacionales 1979–90.

21 See Los Círculos Femeninos Populares. CESAP n.d.a. and n.d.b.

22 The feminist group is only one of the forty-five organizations that are members of CONG. Since the beginning, it has been seen with apprehension by other women: still today, some women are afraid to be stereotyped as 'feminist' and they stress the point that CONG is a 'Confederación de mujeres, no de feministas'.

23 Women from the Círculos Populares seem to have an ambivalent attitude towards CONG: on one hand, they would like to keep their membership in CONG with a low profile because of the fear of being typecast as 'feminist'; on the other hand, they realize how important this membership is to get financing from international agencies who finance women's organizations.

24 I Jornada 'Mujer y Ambiente: La mujer protagonista de la defensa y conservación del ambiente'. Universidad experimental del Tachira UNET. San Cristobal, Venezuela. 1–4 November 1990.

25 The success of the I Jornada 'Mujer y Ambiente' is even greater if we consider that the event took place in the intermediate-size city of San Cristobal, located far from the capital city of Caracas and on the frontier with Colombia.

26 The women organizers estimated that the attendance at this I Jornada was approximately 50 per cent male and 50 per cent female.

27 Interview with organizers and participants in the *I Jornada*: Adicia Castillo, Rosa Trujillo. Caraballeda. 11 January 1992.
28 Interview with Argelia Laya, organizer of the *I Jornada* and Founder of AMAVEN. Caracas, 18 November 1991.
29 Interview with Argelia Laya. Caracas, 11 November 1991.
30 The social relationships between the Executive Committee of AMAVEN are not as strong as in AMIGRANSA or in GEMA.
31 Some members of the Executive Committee reject the possibility of relating the environmental issue with the political parties.
32 The actual President of AMAVEN is also a member of COFEAPRE and participates in CONG through the group 'Area de trabajo hacia la mujer' of the political party Movimiento Hacia el Socialismo, MAS. One of the two founders of GEMA is on the Executive Committee of AMAVEN and some members of COFEAPRE and of CONG are members of AMAVEN.
33 Interview with Giovanna Merola, Caracas, November 1991.
34 Women also participate more than men in government-sponsored organizations that revindicate 'consumption'. More women than men are active in sponsored government Consumers' Associations such as the Asociación de Protección al Consumidor where most of the Presidents appointed by the government have been females.
35 As a consequence of the economic crisis and the macro-structural economic adjustments, Venezuela has changed from a society 'with classes but without social conflict to a society with classes with acute social conflicts'. See Naim and Pioango, 1984.

4 'WE LEARNED TO THINK POLITICALLY'

I am grateful for the full cooperation and involvement of the women participants in the Health Movement of the Jardim Nordeste area, who received me from the outset with warmth, and shared their lives with me, and also the other interviewees who gave me their trust and time. Also, my thanks go to the editors whose comments on an earlier version of this chapter were very constructive.

1 Jardim Nordeste is a district in the Eastern Zone of São Paulo with no autonomy for the administration of healthcare in the area. On the other hand, the Jardim Nordeste area (officially called Ermelino Matarazzo) as described in this chapter is, in fact, a municipal administrative subdistrict for health, part of the overall management structure of healthcare of the State Department of Health of São Paulo. The name of the 'area' in this chapter takes that of the district simply because the Health Movement takes it as well, for it started there with women who lived there.
2 For the antecedents to and the first years of the 1964 military *coup* in Brazil, see Branco 1975 and 1976 respectively.
3 The monthly minimum wage in Brazil was formally established in 1940 at different levels for different regions of the country and did not include expenses relating to education and leisure (Sabóia 1985). In May 1984, however, the minimum wage was unified for the country as a whole (Departamento Intersindical de Estatística e Estudos Sócio-Economicos 1984).
4 São Paulo Metropolitan Area is a conurbation made up of 37 municipalities including the city of São Paulo.
5 Data from EMPLASA 1985: 78.

6 For a study of the Brazilian military regime, see Flynn 1979; Velasco e Cruz and Martins 1983.

7 In Brazil, 'The CEBs are small groups of Christians organized around the local parish by the initiative of clerical or lay people. They first started around 1960 . . . and by 1980 they numbered 80,000 – involving around 1,500,000 people across the country' (Barroso 1982: 154). See also de Camargo *et al.* 1981; Betto 1984.

8 For an account of early feminist struggles in Brazil see Saffioti 1978; Alves 1980; Singer 1981.

9 Data calculated from information in EMPLASA 1985: 127–33.

10 The fieldwork in the Jardim Nordeste area consisted of participant observation from February 1985 to February 1986. All the meetings of the Movement were attended as well as extraordinary activities, such as jumble sales, demonstrations, elections of the Health Council. Also, it included semi-structured interviews and questionnaires, the latter aiming at gathering socio-economic data on the women.

11 The World Health Organization suggests that about four to six beds for every 1,000 inhabitants is a basic minimum.

12 The medical students were at the end of the basic part of their course at the University and starting to get involved with the clinical part of it which included contact with the Clinic's Hospital. This is the main public hospital of the state of São Paulo, where all the students of the medical school of the University of São Paulo, to which it is linked, have to work as part of their course. This contact 'frustrated' them very much because the hospital offered a very biased view of the reality of the health situation since it concentrated more on the rare clinical cases. They then contacted one of their teachers with whom they developed the idea of going out to the outlying and poorest part of the city, to have a closer look at health there. The other motivation, they said, was linked to political participation.

13 For an example of a more direct influence of the Feminist Movement on an urban movement (they discuss the Nursery Movement) see Schmink 1981; Alvarez 1985; Caldeira 1987.

14 SOF is a 'private entity which provides low-income women of both the Southern and Eastern Zones of São Paulo with assistance' (Barroso 1982: 164).

5 WOMEN'S POLITICAL PARTICIPATION IN *COLONIAS POPULARES* IN GUADALAJARA, MEXICO

I would like to thank the ESRC for funding the research for my trip (on which this chapter is based), and also Sallie Westwood, Sarah Radcliffe, Pingla Udit, Citlali Rorirosa Madrazo and Joe Foweraker for their helpful comments.

1 Molyneux describes strategic gender interests as those which are derived from 'the analysis of women's subordination and from the formulation of an alternative more satisfactory set of alternatives' and practical gender interests as arising 'from the concrete conditions of women's positioning within gender division of labour' (Molyneux 1985: 232–3).

2 Une-Ciudadanos en Movimiento is the new name for the CNOP, Confederación Nacional de Organizaciones Populares, one of the three national Confederations. It was reorganized in 1989–90 to try and recoup support for the PRI after the 1988 Federal Elections indicated that the PRI was

under serious threat. For a more detailed account of the reorganization see Craske 1992.

3 *Ejidos* are common lands established after the Revolution. Technically they cannot be built upon but more are becoming engulfed in the urban sprawl. In Mexico City and Monterrey the *ejidos* were invaded and settlements established. In Guadalajara the urbanization of *ejidal* land has been different: in most cases the land has been illegally sold. After this land has been built upon it needs to be 'regularized', which requires payment by the residents to obtain the legal title to the land. In 1975 a government agency, CORETT, was established to deal with land regularization, which in fact was legalizing the illegal (Vázquez 1989b: 105fn).

4 These details were provided by Ricardo Zavala in in his capacity as Director of the Citizens' Participation Office.

5 The 'wearing of two hats' is not uncommon in Mexican urban politics. Bassols and Delgado (1986) comment on a *priísta* leader in a Mexico City *colonia* who is also the secretary for CORETT, the land regularization agency, in the neighbourhood of Azcapotzalco.

6 SEDOC (Educational Services of the East) is a Jesuit organization set up in the 1970s to promote popular education projects and adult literacy programmes in the *colonias populares*. Through this work it became involved with popular mobilizations.

7 It is difficult to know how many people are involved in the PRI and its organizations. In the mid-1980s it claimed to have a membership of 12 million which represented 18 per cent of the total population and around 35 per cent of the adult population (Story 1986: 82). However, this figure probably includes those registered with the groups such as neighbourhood committees, although few of them carry PRI credentials.

8 Many participants are *compadres* to one another through the god-parent relationship. By becoming the god-parent of a child you also strengthen your relationship with the parents.

9 A *caja popular* is a mutual savings bank where a small amount of money is saved each week by the contributors who take it in turn to withdraw the money.

10 Mohr Peterson discusses how women have dealt with objections from their spouses regarding their participation (1990: 22).

11 They are not the only ones to receive such support. There are at least two other organizations in Guadalajara, namely IMDEC and EDOC.

12 Between 1952 and 1988 abstentionism ranged from 26.53 per cent to 48.42 per cent (*La Jornada* 1988) and has generally been higher for Jalisco (Ramos Oranday 1985: 176–4).

13 While I was conducting the fieldwork I was asked by some of the OICO women to help them establish a women's group, the only stipulation being that 'men must not be seen as the enemy'. Thus, they are caught between addressing the problems, without alienating their friends and comrades in the mobilization, and their spouses in their personal relationships.

14 Silvia Hernández is a young woman very much in keeping with the image Salinas wishes to promote. It is worth pointing out that the leader of the Workers Confederation is a veteran of the Revolution. Previously she was Minister of Tourism and also had been involved with youth programmes.

15 Women's activities have been concentrated in the informal politics of neighbourhood and community based politics but they have also become

more involved at the institutional and formal level. On the committee which developed the CNOP/*Une* reorganization, forty-eight of the 246 participants were women, not yet the 30 per cent they claim to want but higher than most such bodies.

16 Many people, particularly from other groups belonging to Intercolonias and OICO, feel the affiliation was pushed by some of the promoters who are officials of the party. These people did not wish to be identified at the present time. The women I interviewed were evasive on the issue; however, the interviews took place a year before the affiliation.

6 FEMALE CONSCIOUSNESS OR FEMINIST CONSCIOUSNESS?

I would like to thank Pepe Roberts, Lynne Brydon and the editors for comments on an earlier version of this chapter.

1 This chapter is dedicated to Elisabeth Souza-Lobo whose untimely death earlier this year represented a significant loss to Brazilian feminist research and to those, like myself, who had the privilege of knowing and working with her.

2 The work of Klaus Offe (1985) on non-institutional politics makes similar omissions with relation to gender.

3 The material on which this chapter is based arises out of a wider research project conducted on the role of women in the organization and formation of popular urban social movements in Brazil during the period 1983–5. The research was funded by a postgraduate award from the Economic and Social Research Council. Fieldwork was carried out in nine low-income neighbourhoods from three different regions of the Greater São Paulo metropolitan area: Embu and Sta Emilia in the Southern Zone; Vila Rica, Vila Antonieta, Vila Sezamo and São Matheus in the Eastern Zone; and Diadema, São Caetano do Sul and Maua in the 'ABC' region to the far east of the metropolitan area. Interviews were held with over 200 women who were active participants in popular movements. All material and quotations used in this chapter are from the author's primary data, unless otherwise stated.

4 Many women faced strong opposition to their political involvement from their partners. In many regions women's self-help groups gave practical help and support to those who were victims of domestic violence or who wished to separate from their partners.

5 The Health Movement in conjunction with feminist groups and women's organizations had strongly opposed all family planning proposals by the government for being far too authoritarian. For example, in 1985 the Head of the Armed Forces insisted that the question of family planning should come under their jurisdiction because it was a matter of National Security!

6 Paulo Freire created a literacy course which utilized political material as subject matter to teach adults, in the space of forty lessons, the basic skills of reading and writing while at the same time developing a political consciousness in the student. In recent years the Brazilian Catholic Church has adapted this method for the conscientization of the popular classes and many political militants attend weekend courses, held by the Church, to learn how to deliver this course.

7 The APSD was one of the successful outcomes of political action by the Unemployed Movement. This Association was sponsored by various ecumenical bodies and Churches in São Paulo to give financial support

to cooperatives and employment schemes initiated by the unemployed themselves.

8 In all the Brazilian states where help was offered by the government, the majority of projects were either submitted by women or were to be carried out by them.

9 Not all the Committees of the Unemployed had cooperative work schemes as part of their organization. Some had allotment schemes which worked on a cooperative basis to grow fruit and vegetables to supplement the diets of the unemployed. These schemes were organized in the same way as other cooperatives, the only difference being that production was for use rather than for sale.

10 People from adjacent neighbourhoods or towns often used *favelas* as dumping grounds for their rubbish. Part of the scheme was to prevent this happening by putting up official notices threatening the culprits with prosecution.

7 TOUCHING THE AIR

1 The translation from the original Spanish is mine. Throughout, unless otherwise stated, all translations are mine, and are intentionally literal translations of the original Spanish. Gabriela Mistral was the first Latin American Nobel Prize Laureate, in 1945.

2 For a history of the fight for the vote see Kirkwood 1990: 90–127.

3 Salvador Allende was President of Chile from 1970–3, at the head of Popular Unity, a democratically elected Marxist government formed by a coalition of left-wing parties. He died during the military *coup* of 11 September 1973, during the takeover of the seat of government, the Moneda Palace, in which General Augusto Pinochet came to power. Pinochet's regime lasted until 1989.

4 Quilapayún was one of the most prominent groups of the *nueva canción chilena*, the new song movement, that used indigenous music, instruments and costume as a way of establishing an 'authentic' Latin American music, free from what was regarded as the cultural imperialism of North American pop. Their music was, at this time, firmly rooted in the politics of Popular Unity, for whom they were 'cultural workers', using song as a way of uniting people around Popular Unity ideology. In the song quoted here, women are exhorted to join the masses breaking the chains of oppression, but are singled out as a group apart from students, employees, workers. The same tendency is seen in the other Quilapayún anthem, 'El pueblo unido jamás será vencido' (The people united will never be defeated), where women are told in the final lines that they also carry justice and right in their hands, and that 'with fire and with courage, woman,/ now you are here,/ side by side with the worker'.

5 See Vodanovic 1970. For an excellent description of Allende's appeals to women see Elsa Chaney 1974, and for a first-hand account of the way women accepted this allotted place, described by Mario Benedetti as '*machismo-leninismo*', see Kirkwood 1990: 49–76.

6 See 'Plebiscite Success Strengthens Pinochet's Personal Position', *Latin American Weekly Report*, 19 September 1980, pp. 6–7.

7 The 'Forty Measures' were the forty points of the Popular Unity programme. See 'En el estadio de Chile: Multitudinario Homenaje de las Mujeres a Pinochet', *El Mercurio*, 18 August 1988. *El Mercurio* gives the number in attendance as 'several thousand women'.

8 'Según Encuesta "Skopus": Mujeres Apoyan en Forma Mayoritaria al Sí', *El Mercurio*, 4 September 1988. Nationwide, the poll showed 41.8 per cent of women in favour of the Yes vote, 29.8 per cent for the No, and 28.4 per cent who did not know. It must be stated, however – and this was not highlighted in the article – that the mixed total in the same poll showed 39.3 per cent for the Yes vote, and 33 per cent for the No. So the actual figures in the poll are not as clear-cut as the headline would have us believe, and the percentage of voters who were still undecided one month before the plebiscite was relatively high in both the female vote and the mixed vote.

9 'A las mujeres chilenas. Demandas de las mujeres a la democracia', *La Epoca*, 1 July 1988.

10 For a comprehensive study of the Popular Unity period and the atmosphere surrounding it see Valenzuela, 1978.

11 For a concise discussion of popular culture see Rowe and Schelling 1991.

12 I have used the word *población* in the original Spanish because I feel that the usual translation of shantytown is inadequate. Here I borrow the definition of Carolyn Lehmann in an as yet unpublished article, 'Bread and Roses: Women Who Live Poverty': 'A *población* is one of the urban, marginated densely populated areas peripheral to the center of the capital city, Santiago. Each *población* has its own character, its own name, its history, its pride.' The word cannot satisfactorily be translated as 'shantytown' or 'urban slum' (pp. 1–2). Some *poblaciones* have well-developed social and cultural organizations.

13 *Arpilleras* are testimonial tapestries made on sackcloth (the Spanish for which is *arpillera*) that depict the lives of people in the *poblaciones* of Chile. Although there is a tradition of embroidery in Isla Negra in Chile, the *arpilleristas* gave this a new form and meaning, for they were the wives and mothers of the disappeared. In their workshops they joined together to share their grief and to express it through these tapestries. They used any scraps of cloth and wool, and sewed the testimony of these years, whose grim reality the regime was trying to make disappear before the eyes of the rest of the population. For further information see Agosín 1987.

14 Taller de Investigación Teatral and David Benavente 1979: 196–248. For an analysis of the play, *Tres Marías y una Rosa*, see Boyle 1986.

15 For a full discussion of women's movements see Kirkwood and Molina 1986.

16 For information on the demonstrations by women throughout the Pinochet regime see Molina 1990.

17 Héctor Noguera, round table discussion, May 1991, Santiago de Chile.

18 Taken from information about the radio station provided by the Casa La Morada.

19 ibid.

20 ibid.

8 ADJUSTMENT FROM BELOW

This chapter was originally written for UNICEF, New York. An earlier, abbreviated version, entitled *The Impact of Recession and Adjustment Policies at the Micro-Level: Low Income Women and their Households in Guayaquil, Ecuador*, was published in UNICEF (1989). The 1988 fieldwork and analysis of research data was undertaken in collaboration with Peter Sollis. I would like to acknowledge his important contribution to this chapter. I would also like to thank Michael Cohen for encouraging me to examine adjustment processes at the micro-level; Richard Jolly for his interest in this research, and Diane Elson for her helpful comments on this chapter. Without the support and commitment of the *moradores* of Indio Guayas it would not have been possible to do this research. As ever I am indebted to them, particularly to Emma Torres, Rosa Vera and Lucie Savalla, who after ten years are much better fieldworkers in their own community than I could ever be.

1 To assist the formulation of policy, the World Bank, in collaboration with UNDP and the African Development Bank, have embarked on an extensive research programme to monitor the effects of SAPs in sub-Saharan Africa, setting up the SDA Project Unit, and undertaking the Permanent Household Survey (PHS), a large-scale structured household interview survey to be undertaken over a five to seven year period (UNDP 1987). At the same time a proliferating number of detailed micro-level studies of the effects of SAPs on women have been commissioned (Commonwealth Secretariat 1990).

2 Fieldwork for this longitudinal study was first undertaken in 1977–8, and was based on participant observation through living in the *barrio*, and a survey of 244 households of three different block groups, intended to show changes in settlement and consolidation processes. Further anthropological fieldwork was undertaken in January 1979 and August 1982. A restudy of the *barrio* was undertaken in July–August 1988, based once more on participant observation, and a sample survey of 141 households in the same area (referred to as the survey). A further semi-structured questionnaire was undertaken with a sub-sample of 33 households from the sample survey, selected to be representative of the different household structures in Indio Guayas (referred to as the sub-sample). The purpose of the sub-sample was to examine in greater depth at a qualitative level important processes highlighted in the survey. Analysis of the issues therefore relates only up to the period of August 1988, and not to further changes which may have occurred since then. For a more detailed description of the research methodology see Sollis and Moser 1990.

3 Elaborating on this issue Elson argues that,

> taking no account of gender leads to the belief, expressed by the Chief of the Trade and Adjustment Policy Division in the World Bank, that 'it is relatively easy to retrain and transfer labour originally working in, say, construction or commerce, for employment in the export . . . of, say, radios or garments'
>
> (Selowsky 1987) (Elson 1990: 8)

4 The changes identified as of greatest importance were selected after useful discussions with Diane Elson and Francis Stewart. In fact, for policy makers the most important purpose of research such as this relates to the issue of causality, and the identification as to which of the social costs experienced are a consequence of debt or recession rather than of the IMF/World Bank stabilization/SAP interventions. Of equal importance for them is the question of counterfactuals, and the thorny problem as to what conditions would have been like, had SAPs not occurred. At the outset it is critical to recognize the difficult methodological problems in directly answering questions such as these and therefore the necessity of identifying changes resulting from both recession and adjustment generally, with more specific inferences drawn out wherever possible.

5 During the past twenty-five years Ecuador has experienced average annual GDP growth of 6.7 per cent, although from 1982 to 1988 the rate was only

2.2 per cent per annum. The fluctuating rates reflect the development and decline of the petroleum and oil industry. Oil exports rose from 0 per cent to 70 per cent of total exports between 1970 and 1983 although with the subsequent decline in oil prices since then its share has dropped to about 42 per cent of total exports, below agriculture at 52 per cent.

6 Both these figures are much higher than the national average which in 1982 was 18.55 per cent. The differences result not only from the fact that the Indio Guayas sample was biased towards a low-income population but also because of the 'invisibility' of so much of women's productive work in national accounting.

7 Although increasing differentiation may continue among the next generation of women, employment opportunities were not commensurate with the increased educational qualifications, and the likelihood is that the majority, once they have their own dependants, will retreat into the female occupations of selling and domestic service.

8 I am grateful to Diego Carrión for his analysis of this issue.

9 One resourceful woman, whose husband would not give permission, persuaded the hospital to accept instead the permission of her mother and her brother, arguing that the two together were equal to her husband.

10 These provisional estimates as to the size of each group have been made on the basis of conclusions reached as a result of both qualitative and quantitative research undertaken in August 1988.

9 'PEOPLE HAVE TO RISE UP – LIKE THE GREAT WOMEN FIGHTERS'

Many thanks are due to the Nuffield Foundation for funding research in Peru on two occasions, as well as to the Centre of Latin American Studies, University of Cambridge, for time to write and think. I am very grateful to Janett Vengoa for her transcription and translation work. Thanks are also due to Sallie Westwood, Felix Driver, Peter Jackson, Alison Scott and Sarah Skar for helpful comments on an earlier draft.

1 Political culture is defined as a domain 'where meanings are negotiated and relations of dominance and subordination are defined and contested' (Jackson 1989: 2). Closely linked to political culture, when discussing low-income and mostly powerless groups is the notion of 'popular culture'. Popular culture refers to 'a space or series of spaces where popular subjects, as distinct from members of ruling groups, are formed' (Rowe and Schelling 1991: 10).

2 Previous work by geographers on the spatial variation in gender roles and gendered relations of politics have focused on First World societies (WGSG 1984). Therefore, much work remains to be done on the geographical specificities of gender and politics, in terms of identities and relations in Third World areas, in order to clarify the ways in which space constitutes and defines male and female political roles (but see *Political Geography Quarterly* 1990; Radcliffe 1990a, 1990b).

3 The state is defined here as a non-unitary machinery for the exercise of government over a given population (usually defined spatially), and as constituting a body of institutions organized around an intention to control. States have an apparatus of enforcement and act through a combination of ideological, judicial and repressive processes (Anthias and Yuval-Davis 1989: 4).

231

Moreover, states are based upon particular configurations of ethnic, class and gender relations which provide project(s) for governance (Foucault 1980).

4 However, an emphasis on the state should not blind us to the analysis of other spatial and social arenas in which femininities (and masculinities) are being formed (Radcliffe 1990a). Regional, local and domestic activities and contestations are also important in the configuration of political behaviour (Foucault 1980).

5 The gender regime is defined by Bob Connell (1987: 98–9) as the 'structural inventory of a historically constructed pattern of power relations between men and women and definitions of femininity and masculinity of a particular institution'.

6 Debate continues over the extent to which gender issues (and particularly women's position) is central to and/or explicit in the state's project (see Anthias and Yuval-Davis 1989: 1; MacKinnon 1982; Alvarez 1990; Franzway et al. 1989).

7 Following from this, is the question of where state-sanctioned gender identities originate. I suggest that specific gendered identities arise from pre-existing social relations, subsequently elaborated and shaped by state institutions and political culture (see Moore 1988: 128–9).

8 The incorporation of certain masculinities and femininities into the power apparatus of the state has been addressed by other analysts of Latin American politics, although not using these terms. Elsa Chaney (1979), in her study of women politicians in Chile and Peru, emphasized conformity to a particular femininity around an elite, Catholic, European norm. Identities were associated with the moral superiority and incorruptibility of 'Woman', compared with masculine politics which drew on metaphors of ruthless behaviour and corruption.

9 Population and fertility policies of the Peruvian state are not addressed here, as this theme has been highlighted elsewhere (Afshar 1987; Anthias and Yuval-Davis 1989). Moreover, issues around the body were not directly linked to political identities among Peruvian peasants until the civil war between guerrillas and army in the 1980s, when women began to organize into groups for 'the disappeared', and complain of rapes of peasant women by soldiers (on Bolivia see Sostres and Ardaya 1984: 17).

10 Previous to 1974, the only government institution to deal with women was the Junta de Asistencia Nacional (JAN: National Assistance Council), under the presidency of the President's wife. JAN was created in the 1950s and was substantially influenced by Eva Peron's role in Argentina (Portocarrero 1990: 112).

11 Quechua is used in this context to refer to highland indigenous peasantries in Peru, most of whom are found in the southern Andes. It is estimated that some 9 million people speak Quechua, the language utilized and spread by the Inca empire.

12 This name can also be spelt Bastídas.

13 For a history highlighting Micaela Bastídes' role, see Prieto 1980.

14 These feminist groups included Acción para la Liberación de la Mujer Peruana (ALIMUPER) Promoción de la Mujer and Grupo de Trabajo Flora Tristan, later linked in the Frente de Mujeres (Ruiz Bravo 1987: 107). However, the impact of feminist groups on state constitution of masculinities or femininities was minimal as they had no government or official posts. Moreover, the women involved in these groups were almost exclusively

Limeña, middle-class educated women, remote from *campesina* and shanty-town women's experience (Vargas 1986).

15 ACOMUC was one of a series of institutions set up by the Velasco government to encourage female participation in 'society'. Others included Comité Técnico de Revaloración de la Mujer (COTREM: Technical Committee for the Revalorization of Women, founded 1973), Comisión Nacional de la Mujer Peruana (CONAMUP: National Council of Peruvian Women) which drafted proposals for work legislation, researched prostitution, and carried out training, and Instituto Nacional de Protección al Menor y la Familia (INAPROMEF: National Institute for the Protection of Minors and the Family) (Ruiz Bravo 1987; Portocarrero 1990).

16 'Mothering' politics, defined as valuing and elevating the traditional role of motherhood, is distinct from 'motherist' politics, that is women themselves using historical images of motherhood as subversive symbols in dealing with the state. The term 'motherist' as a term for women's groups was first used in relation to the groups of Mothers of the Disappeared in Argentina, Guatemala and Chile (see Schirmer 1989).

17 However, the creation in 1983 of the Oficina de la Mujer (Office of Women) in the Justice Ministry was a contradictory element in that it had a remit to coordinate all government actions in favour of women. In effect it was underfunded and politically weak and had little impact on the shape of state femininities (Portocarrero 1990: 114–15).

18 The Programa de Apoyo de Ingreso Temporal (PAIT: Temporary Income Support Programme) was a public works programme granting work for three months at a time and paid at the legal minimum wage rate (approx. US\$36 per month). Three-quarters of the workers were women, although the programme was not designed explicitly for women. (For excellent discussions of PAIT see Graham 1991; Paredes and Tello 1988).

19 Peasant women have been identified by initials only, to retain confidentiality.

20 On the other hand, the richest 20 per cent received 60.3 per cent of national income (*Situación Latinoamericana* 1991: 159).

21 Data for this section is derived from interviews with female and male national and regional leaders of the two major peasant unions, CCP and CNA, in 1988 (July–September) and 1990 (January–March). Questionnaires were followed in one-third of cases by semi-structured interviews, which were taped and then transcribed. Secondary sources, such as union publications and published interviews with confederation union leaders, were also consulted.

22 Their femininity at this time was also being renegotiated in the domestic space, with families and spouses (see Radcliffe forthcoming).

23 For a comparison on Mexico, see Stephen 1989.

24 Clorinda Matto de Turner (1854–1909) was a writer (she wrote novels dealing with the abuses of Indians by landowners), and editor of the daily *La Bolsa* and other newspapers.

25 As noted above, the situation of other low-income women was not so secure and they could not always prioritize their own self-representation in relation to APRA Mothers' Clubs and work creation projects.

26 However, given the political and economic insecurity prevailing in the country during the late 1980s and early 1990s, they were unable to carry out meetings on the scale originally envisaged: by early 1990, the Committee had all but abandoned large meetings.

27 The global symbolic importance of International Women's Day on 8 March is highlighted by this name.

28 'Although men and women have distinct ways of being, thinking, feeling and seeing things, ... both have the same value and importance for God and deserve equal respect' (Instituto de Estudios Aymaras 1985).
29 More work is needed on the construction of masculinities during this same period in Peru, with the continuous interaction between masculinities and femininities in creating the gender regime (see Scott 1990 for a discussion of masculinities among Peruvian urban low-income groups). Moreover, implicit state notions of appropriate masculinities affect policies and programmes (see Westwood 1991; Jackson 1991 on the British context).

REFERENCES

1 GENDER, RACISM AND THE POLITICS OF IDENTITIES IN LATIN AMERICA

Alvarez, S. (1986) *The politics of gender in Latin America*, unpublished PhD thesis: Yale University.
—— (1990) *Engendering Democracy in Brazil*, Princeton NJ: Princeton University Press.
Alves, C. S. (1990) 'The Brazilian economic crisis and its impact on the lives of women', *Political Geography Quarterly* 9(4): 415–25.
Anderson, B. (1983) *Imagined Communities: Reflections on the Origin and Spread of Nationalism*, London: Verso.
Andreas, C. (1985) *When Women Rebel: The Rise of Popular Feminism in Peru*, Westport, Conn.: Lawrence Hill.
Armstrong, W. and McGee, T. (1985) *Theatres of Accumulation: Studies in Asian and Latin American Urbanisation*, London: Methuen.
Ardener, S. (ed.) (1981) *Women and Space: Ground Rules and Social Maps*, London: Croom Helm.
Ballón, E. (ed.) (1986) *Movimientos Sociales y Democracia: La Fundación de un Nuevo Orden*, Lima: DESCO.
Barrios de Chungara, D. (1978) *Let Me Speak! Testimony of Domitila, a woman of the Bolivian Mines*, Monthly Review Press: New York.
Bhabha, H. K. (1983) 'The other question', *Screen*.
Blondet, C. (1986) *Muchas vidas construyendo una identidad: mujeres pobladoras de un barrio limeño*. Working paper no. 9, Lima: Instituto de Estudios Peruanos.
Bobbio, N. (1989) 'Gramsci and the concept of civil society', in J. Keane (ed.) *Civil Society and the State*, London: Verso: 73–100.
Bourque, S. (1989) 'Gender and the state: perspectives from Latin America', in S. Charlton, J. Everett, and K. Staudt (eds) *Women, the State and Development*, Albany: SUNY.
—— and Warren, K. (1989) 'Democracy without peace: the cultural politics of terror in Peru', *Latin American Research Review* 24(1): 7–34.
Bourricaud, F. (1975) 'Indian, *mestizo* and *cholo* as symbols of the Peruvian system of stratification', in N. Glazer and D. Moynihan (eds) *Ethnicity: Theory and Experience*, Cambridge, MA: Harvard University Press.
Boyne, R. and Rattansi, A. (eds) (1990) *Postmodernism and Society*, London: Macmillan.
Bunster, X. (1980) 'Surviving beyond fear: women and torture in Latin America', in H. Safa and J. Nash (eds) *Women and Change in Latin America*, South Hadley, MA: Bergin and Garvey.

REFERENCES

Butler Flora, C. (1982) *Socialist Feminism in Latin America*: Working Paper 14 W1D, East Lansing: Michigan State University.

Cahiers des Ameriques Latines (1982) 'Mouvements des femmes en Amerique Latine' 26, Paris: Sorbonne.

Castells, M. (1978) *City Class and Power*, London: Macmillan.

Chaney, E. (1979) *Supermadre: Women in Politics in Latin America*, Austin: Institute of Latin American Studies, University of Texas Press.

Chile: Women's Delegation Report (1986) *Somos Más: Gathering Strength*, London: Chile Solidarity Campaign.

Chinchilla, N. (1979) 'Working-class feminism: Domitila and the housewives committee', *Latin American Perspectives* 6(3): 87–92.

Collier, D. (ed.) (1979) *The New Authoritarianism in Latin America*, Princeton, NJ: Princeton University Press.

Collinson, H. (ed.) (1990) *Women and Revolution in Nicaragua*, London: Zed.

de Lauretis, T. (1987) *Technologies of Gender*, London: Macmillan.

Diamond, I. and Quinby, L. (eds) (1988) *Feminism and Foucault: Reflections on Resistance*, Boston: Northeastern University Press.

Driver, F. (1985) 'Power, space and the body: a critical assessment of Foucault's *Discipline and Punish*', *Environment and Planning D: Society and Space* 3: 425–46.

Eckstein, S. (ed.) (1986) *Power and Popular Protest: Latin American Social Movements*, Berkeley: University of California Press.

Foucault, M. (1980) *Power/Knowledge: Selected Interviews and Other Writings 1972–1977* Brighton: Harvester Press.

Fraser, N. and Nicholson, L. (1988) 'Social criticism without philosophy: An encounter between feminism and postmodernism', *Theory Culture and Society* (2–3): 373–94.

Gledhill, J. (1988) 'Agrarian social movements and forms of consciousness', *Bulletin of Latin American Research* 7(2): 257–76.

Gonzales, L. (1988) *Women Organising for Change: Confronting the Crisis in Latin America*, Rome/Santiago: Isis International in coordination with DAWN.

Hall, S. (1987) 'Minimal selves' in ICA Document 6, London: Institute of Contemporary Arts.

—— (1991) 'Europe's other self', *Marxism Today* August 18–19.

Howard-Malverde, R. (1990) *The Speaking of History: 'Willapaakushayki' or Quechua Ways of Telling the Past*, Research paper no. 14, London: Institute of Latin American Studies.

ISIS (1986) *The Latin American Women's Movement: Reflections and Actions*, ISIS Women's Journal 5, Santiago/Rome: ISIS.

Inden, R. (1990) *Imagining India*, Oxford: Basil Blackwell.

Jaquette, J. (1973) 'Women in revolutionary movements in Latin America', *Journal of Marriage and the Family* 35(2): 344–54.

—— (ed.) (1989) *The Women's Movement in Latin America: Feminism and the Transition to Democracy*, London: Unwin Hyman.

Jelin, E. (ed.) (1990) *Women and Social Change in Latin America*, London: Zed/UNRISD.

Kaplan, T. (1982) 'Female consciousness and collective action: the case of Barcelona, 1910–1918' *Signs* 7(3): 545–66.

Labao Reif, L. (1989) 'Women in revolution: the mobilization of Latin American women in revolutionary guerrilla movements', in L. Richardson, V. Taylor (eds) *Feminist Frontiers II*, New York: Random House.

Laclau, E. (1987) (1990) *New Reflections on the Revolution of Our Time*, London: Verso.

236

REFERENCES

Laclau, E. and Mouffe, C. (1985) *Hegemony and Socialist Strategy: Towards a Radical Democratic Politics*, London: Verso.
LADOC (1984) 'Grandmothers of the Plaza de Mayo: sorrow and hope', *LADOC Newsletter* XV(26): 29–37.
Latin American Perspectives (LAP) (1991) Issue on Testimonies in Latin America. 70(18): 3.
Lewis, P. (1971) 'The female vote in Argentina, 1958–65', *Journal of Comparative Political Studies* 3: 425–41.
Logan, K. (1989) 'Empowerment within a female consciousness: collective actions in Mérida, Yucatan, Mexico', paper given at Latin American Studies Association XV International Congress, Puerto Rico, September.
Long, N. and Villarreal, M. (1989) 'The changing life-worlds of women in a Mexican *ejido*: the case of the beekeepers of Ayuquila and the issue of intervention', in N. Long (ed.) *Encounters at the Interface*, Wagenigen: Wagenigen Agricultural University.
MacDonald, N. (1991) *Brazil: A Mask Called Progress*, Oxford: OXFAM.
Macías, A. (1982) *Against All Odds: the Feminist Movement in Mexico to 1940*, Westport: Greenwood Press.
Malloy, J. M. and Seligson, M. (eds) (1987) *Authoritarians and Democrats: Regime Transition in Latin America*, Pittsburgh: Pittsburgh University Press.
Mani, L. (1984) 'The production of an official discourse on SATI in early nineteenth-century Bengal' in F. Barker *et al.* (eds) *Europe and Its Others*, Wivenhoe: University of Essex.
McClintock, C. and Lowenthal, A. (eds) (1984) *The Peruvian Experiment Reconsidered*, Princeton, NJ: Princeton University Press.
McGee, S. (ed.) (1981) 'Women and politics in 20th Century Latin America' *Studies in Third World Societies* 15.
Menchú, R. (1983) *I, Rigoberta Menchú: an Indian Woman in Guatemala*, London: Verso.
Mohanty Talpade, C. (1988) 'Under Western eyes, feminist scholarship and colonial discourses', *Feminist Review* 30: 61–88.
Molyneux, M. (1985) 'Mobilisation without emancipation? Women's interests, the state and revolution in Nicaragua', *Feminist Studies* 11(2): 227–54.
Munck, R. (1990) *Latin America: The Transition To Democracy*, London: Zed.
Nascimento, do A. (1989) *Brazil: Mixture or Massacre: Essays in the Genocide of Black People*, Dover, MA: Majority Press.
Navarro, M. (1989) 'The personal is political: Las madres de Plaza de Mayo', in S. Eckstein (ed.) *Power and Popular Protest: Latin American Social Movements* London: University of California Press.
O'Brien, P. and Cammack, P. (eds) (1985) *Generals in Retreat*, Manchester: Manchester University Press.
O'Donnell, G. (1973) *Modernisation and Bureaucratic Authoritarianism: Studies in South American Politics*, Berkeley: University of California Press.
O'Donnell, G., Schmitter, P. and Whitehead, L. (eds) (1986) *Transitions From Authoritarian Rule*, Baltimore: Johns Hopkins University Press.
O'Hanlon, R. (1988) 'Recovering the subject: subaltern studies and histories of resistance in colonial South Asia', *Modern Asian Studies* 22(1): 61–79.
Patai, D. (1988) 'Constructing a self: a Brazilian life story', *Feminist Studies* 14(1): 143–66.
Philip, G. (1978) *The Rise and Fall of the Peruvian Military Radicals, 1968–1976*, London: Athlone Press.

REFERENCES

Prieto, M., Rojo, C. *et al.* (1987) *No sé quién nos irá a apoyar: el voto de la ecuatoriana en mayo de 1984*, Quito: CEPLAES. Cuadernos de la mujer 4.

Radcliffe, S. (1990) 'Multiple identities and negotiation over gender: female peasant union leaders in Peru', *Bulletin of Latin American Research* 9(2): 229–47.

—— (in press) *Confederations of Gender: The Mobilisation of Peasant Women in Peru*, Ann Arbor: University of Michigan Press.

Redclift, M. (1988) 'Introduction: agrarian social movements in contemporary Mexico', *Bulletin of Latin America Research* 7(2): 249–56.

Rodriguez Villamil, Silvia (1984) 'Los movimientos sociales en la transición a la democracia', mimeo. Montevideo (quoted in Jelin 1990).

Roseberry, W. (1989) *Anthropologies and Histories*, Brunswick and London: Rutgers University Press.

Rowe, W. and Schelling, V. (1991) *Memory and Modernity: Popular Culture in Latin America*, London: Verso.

Said, E. W. (1978) *Orientalism*, Harmondsworth: Penguin.

Sarti, C. (1989) 'The panorama of feminism in Brazil', *New Left Review* 173: 75–92.

Schirmer, J. G. (1989) '"Those who die for life cannot be called dead": Women and human rights protest in Latin America', *Feminist Review* 32: 3–29.

Schmidt, S. (1976) 'Political participation and development: the role of women in Latin America', *Journal of International Affairs* 30: 257–70.

Schmink, M. (1986) 'Women in Brazilian "Abertura" politics', in L. Daube, E. Leacock and S. Ardener (eds) *Visibility and Power*, Delhi: OUP.

Seed, P. (1991) '"Failing to Marvel" Atahualpa's Encounter with the Word', *Latin American Research Review* 26(1): 7–32.

Slater, D. (1985) *New Social Movements and the State in Latin America*, CEDLA Latin American Studies 25, Holland: Foris Publications.

Stark, L. (1981) 'Folk models of stratification and ethnicity in the highlands of modern Ecuador', in N. Whitten (ed.) *Cultural Transformations and Ethnicity in Modern Ecuador*, Urbana: University of Illinois Press.

Stevens, E. P. (1973) '*Marianismo*: the other face of machismo in Latin America', in A. Pescatello (ed.) *Female and Male in Latin America*, London and Pittsburgh: University of Pittsburgh Press.

'Talitha Cumi' (1984) *Talitha Cumi circle of Christian feminists: reflections on the theology of liberation*, Lima: 'Talitha Cumi'.

Tovar, T. (1986) '*Barrios, ciudad, democracia y política*' in E. Ballón (ed.) *Movimientos Sociales y Democracia: Fundación de un Nuevo Orden*, Lima: DESCO.

van den Berghe, P. and Primov, G. (1977) *Inequality in the Peruvian Andes: Class and Ethnicity in Cuzco*, Columbia, Mo.: University of Missouri Press.

Vargas, V. (1990) *The Women's Movement in Peru: Rebellion into Action*, WP H12 155, The Hague: Institute of Social Studies.

Viezzer, M. (1979) '*El comité de amas de casa del Siglo XX:* an organizational experience of Bolivian women', *Latin American Perspectives* 6(3).

Villareal, M. (1990) 'A battle over images: women beekeepers' struggles for identity in an *ejido* context', paper given at the Society of Latin American Studies Annual Conference, Oxford, April.

Wachtel, N. (1977) *The Vision of the Vanquished*, Brighton: Harvester Press.

Wade, P. (1986) 'Patterns of race in Colombia', *Bulletin of Latin American Research* 5(2): 1–19.

Westwood, S. (1990) 'Racism, black masculinity and the politics of space', in

238

J. Hearn and D. Morgan (eds) *Men, Masculinity and Social Theory*, London: Unwin-Heinemann.

—— (1991) 'Red star over Leicester: racism, the politics of identity and black youth in Britain', in P. Webner and M. Anwar (eds) *Black and Ethnic Leadership in Britain*, London: Routledge.

Wilson, F. (1986) 'Conflict on a Peruvian hacienda', *Bulletin of Latin American Research* 5(1): 65–94.

Winant, H. (1992) 'Rethinking Race in Brazil', *Journal of Latin American Studies* 24(1): 173–92.

Yuval-Davis, N. and Anthias, F. (1989) *Woman–Nation–State*, London: Macmillan.

Zabaleta, M. (1986) 'Research on Latin American women: in search of our political independence', *Bulletin of Latin American Research* 5(2): 97–103.

2 THE SEEKING OF TRUTH AND THE GENDERING OF CONSCIOUSNESS

Americas Watch (1986) *Civil Patrols in Guatemala*, New York, August.

—— (1989) *Persecuting Human Rights Monitors. The CERJ in Guatemala*, New York, May.

—— (1991) *Guatemala. Getting Away with Murder*, New York, August.

Amnesty International (1991) *Lack of Investigating Human Rights Abuses: Clandestine Cemeteries*, London: AI, March.

Anonymous (1988) 'New widows' group forms', *Central America Report* 23, September: 292.

Armstrong, R. and Shenck, J. (1982) *El Salvador. The Face of Revolution*, Boston: South End Press.

Black, G., Jamail, M. and Chinchilla, N. (1984) *Garrison Guatemala*, New York: Monthly Review Press.

CoMadres (1977–91) Bulletins, Press Releases, etc. (from the FEDEFAM office in Caracas).

CONAVIGUA (1988) *Statement of Founding of CONAVIGUA* (translated into English) 12 September.

—— (1989) 'En El Día Internacional de la Mujer Saludamos a las Mujeres del Mundo y a Nuestras Hermanas Guatemaltecas', *Campo Pagado*, Prensa Libre 8 March.

—— (1991a) *La voz de los Niños Pobres de Guatemala*, Boletín 2. March 1991.

—— (1991b) *Ponencia de la Coordinadora Nacional de Viudas de Guatemala Ante la I. Conferencia de Sectores Damnificados por la Represión y la Impunidad.* Guatemala: 18 and 19 July.

Consejo de Comunidades Etnicas Runujel Junam (CERJ) (1991) *Declaración del Consejo de Comunidades Etnicas: La Historia de los Miembros del CERJ Asesinados y Desaparecidos*, Santa Cruz del Quiche, Guatemala: June.

Elshtain, J. B. (1982) *'Antigone's Daughters' in Public Man, Private Woman: Women in Social and Political Thought*, Princeton, NJ: Princeton University Press: 46–59.

—— (1987) *Women and War*, New York: Basic Books.

—— (1990) 'The Problem with Peace', in J. B. Elshtain and S. Tobias (eds) *Women, Militarism and War*, Savage, Md: Rowman and Littlefield.

Foucault, M. (1977) *'Nietzsche, Genealogy, History'* in *Language, Counter-Memory, Practice: Selected Essays and Interviews*, trans. D. F. Bouchard and S. Simon (ed.), Bouchard, New York: Ithaca.

Frankel, A. (1990) 'Weeping widows no longer: women organise in Guatemala', *Central America Report*, Winter: 6–7.

Fried, J., Gettleman, M., Levenson, D. T. and Peckenham, N. (1983) *Guatemala in Rebellion. Unfinished History*, New York: Grove Press.

de Grazia, V. (1987) 'Heartless havens', *The Nation*, 18 April.

Hawkesworth, M. E. (1989) 'Knowers, knowing, known: feminist theory and claims of truth', *Signs* 14(3): 533–57.

Kaplan, T. (1982) 'Female consciousness and collective action: the case of Barcelona 1910–1918', *Signs* 7(3): 545–66.

Koonz, C. (1987) *Mothers in the Fatherland: Women, Family and Nazi Politics*, London: St Martin's Press.

McClintock, M. (1985a) *The American Connection. State Terror and Popular Resistance in El Salvador*. Vol. I, London: Zed Books.

—— (1985b) *The American Connection. State Terror and Popular Resistance in Guatemala*.Vol. II, London: Zed Books.

Molyneux, M. (1985) 'Mobilization without emancipation? Women's interests, the state and the revolution in Nicaragua', *Feminist Studies* 11(2): 227–54.

Riley, D. (1988) *'Am I That Name?' Feminism and the Category of Women in History*, London: Macmillan.

Ruddick, S. (1984) 'Maternal thinking', in J. Treblicot (ed.) *Mothering, Essays in Feminist Thinking*, Totowa, NJ: Rowman and Allanheld.

—— (1990) 'The Rationality of Care', in J. B. Elshtain and S. Tobias (eds) *Women, Militarism and War*, Savage, Md: Rowman and Littlefield.

Schirmer, J. (1988) '"Those who die for life cannot be called dead": women and human rights protest', *Harvard Human Rights Yearbook*, Inaugural Issue, April, (reprinted in *Feminist Review*, London, autumn 1989).

—— (1991) 'The Guatemalan military project: an interview with Gen. Hector Gramajo', *Harvard International Review* 13(3): 10–13.

Scott, J. W. (1991) 'The Evidence of Experience', *Critical Inquiry* 17: 773–97.

3 *ECOLOGIA*: WOMEN, ENVIRONMENT AND POLITICS IN VENEZUELA

Abramovitz, J. and Nichols, R. (1992) 'Women and biodiversity: ancient reality, modern imperative, development', *Journal of the Society for International Development*, no. 1992: 2 (in press).

Alvarez, M. del Mar (1975) *El feminismo. Espresión de una rebeldía*, Caracas: UCV, Facultad de Ciencias Económicas y Sociales (Manuscript).

Alvarez, S. (1989) 'Women's movements and gender politics in the Brazilian transition', in J. Jaquette (ed.) *The Women's Movement in Latin America*, Boston: Unwin Hyman.

AMAVEN (1991) *Estatutos de la Asociación Mujer y Ambiente de Venezuela*, Caracas: AMAVEN.

AMIGRANSA (1989) *Boletín de los Grupos Ambientalistas Venezolanos*, Renacuajo, no. 1.

AMIGRANSA (1991) *Boletín de los Grupos Ambientalistas*, Renacuajo no. 4, Caracas.

Anonymous (1991) *El Rol de la Mujer en el Desarrollo Sostenible: National Report*. Working paper presented to UNECD, Caracas.

Bartra, E. (1983) *La revuelta*, Mexico: Martin Casilla.

Bennet, V. (1989) 'Urban public services and social conflict: water in Monterrey', in A. Gilbert (ed.) *Housing and Land in Urban Mexico*, Monograph Series, 31, La Jolla: Center for US–Mexican Studies, University of California, San Diego.

Boletín de los Grupos Ambientalistas, Renacuajo. Nos 1, 2, 3, 4 & 5. Caracas.

REFERENCES

Boserup, E. (1970) *Women's Role in Economic Development*, London: George, Allen and Unwin.

Castells, M. (1983) *The City and the Grassroots*, London: Edward Arnold.

CONG (1988) *Registro Civil de la Coordinadora de Organizaciones no Gubernamentales de Mujeres*, 4 July, Caracas: CONG.

—— (1991) *Coordinadora de Organizaciones No Gubernamentales de Mujeres*, Folleto.

CENDES (1989) *Cuadernos del CENDES* 10 (Special Issue: 27/28 February), Centro de Estudios del Desarrollo, CENDES.

CESAP (n.d.a) *¿Qué son los Círculos Femeninos?* Caracas: CESAP.

—— (n.d.b) *¿Cómo Formamos los Círculos Femeninas Populares?* Caracas: CESAP.

Congreso de Asamblea Global de Mujer y Ambiente (1989) *Primera Declaración de las Mujeres de la Región Latinoamericana y del Caribe*, 4–8 November.

Consulta Interamericana sobre Mujer y Medio Ambiente: Estrategias dentro del contexto de las cuestiones ambientales y el desarrollo sostenible (1991) *Declaración y Conclusiones*.

De la Cruz, R. (1988) *Venezuela en Busca 'de un Nuevo Pacto Social*, Caracas: Alfadil/Trópicos.

Donda, F., Grupo Feminista Miércoles. Various interviews through the years 1978–91. Caracas.

Encuentros Feministas Nacionales (1979–90) *Conclusiones*, Venezuela.

Espina, T. and Patiño, M. (1984) *Representación Social del Feminismo*. Caracas: UCV, Escuela de Psicología (Manuscript).

FORJA (1988–91) *Balances Ambientalistas Anuales*, Caracas.

Fourth World Congress on National Parks and Protected Areas (1990) Caracas. 10–17 February.

Foweraker, J. and Craig, A. (1990) *Popular Movements and Political Change in Mexico*, Boulder and London: Lynne Reinner Publishers.

Friedman, J. (1961) *Regional Development Policy: A Case Study of Venezuela*, Boston: The MIT Press.

García Guadilla, M. P. (1981) 'La modernización y el conflicto de roles en la mujer venezolana', *Ponencia presentada en la Reunión del Grupo Santa Lucia*, Bermudas.

—— (1986) 'La experiencia venezolana con los polos de desarrollo: ¿un fracaso del modelo teórico, de la institución planificadora o del estilo de planificación?'. *Cuadernos de la Sociedad Venezolana de Planificación*, 162.

—— (1987) 'Impactos socio-económicos, políticos y espaciales de las grandes inversiones minero-industriales en América Latina: aproximación teórico-metodológica', *Revista Interamericana de Planificación*, 21(81).

—— (1988a) 'Viabilidad política de las demandas y propuestas ambientalistas del Estado y la Sociedad Civil en Venezuela', *Instituto Latinoamericano de Investigaciones Sociales*, Caracas: ILDIS.

—— (1988b) *Directorio de Organizaciones Ambientalistas*, Caracas: ILDIS.

—— (1989) 'Crisis y conflictos socioeconómicos en la Venezuela post-saudita: hacia una redefinición de los actores, los roles y las demandas sociales', in *Consecuencias Regionales de la Reestructuración de los Mercados Mundiales*, Centro de Estudios Urbanos CEUR/Ebert Foundation.

—— (1990) 'Actores y movimientos sociales en los grandes proyectos de inversión minero-industriales en América Latina: Hipótesis sobre la estructuración de la organización social', *Revista Interamericana de Planificación* 23(89).

—— (1991a) 'Crisis, estado y sociedad civil: conflictos socio-ambientales en Venezuela', in García Guadilla, María-Pilar (ed.) *Ambiente, Estado y*

REFERENCES

Sociedad: Crisis y, Conflictos Socioambientales en América Latina y Venezuela. Caracas Universidad Simón Bolívar and Centro de Estudios del Desarrollo, CENDES.

—— (1991b) 'Venezuela: La red de organizaciones ambientalistas', in *El Nudo de la Red: Revista Cultural de Movimientos Sociales*, 16.

—— (ed.) (1991c) *Ambiente, Estado y Sociedad: Crisis y Conflictos Socioambientales en América Latina y Venezuela*, Caracas: Universidad Simón Bolívar and Centro de Estudios del Desarrollo, CENDES.

—— (1992) 'Symbolic effectiveness, social practices, and strategies of the Venezuelan environmental movement: its impact on democracy', in A. Escobar and S. Alvarez (eds) *New Social Movements in Latin America: Identity, Strategy and Democracy*, Westview Press.

Gerez-Fernandez, P. (1990) *Movimientos y Luchas Ecológicas en México* (Manuscript).

Gomez, L. (ed.) (1987) *Crisis y Movimientos Sociales en Venezuela*, Caracas: Trópicos.

—— (1991) 'Estado, ambiente y sociedad civil en Venezuela: convergencias y divergencias', in M. P. García-Guadilla. (ed.) *Ambiente, Estado y Sociedad: Crisis y Conflictos Socioambientales en América Latina y Venezuela*.

Grupo Ambientalista ECO XXI (1987, 1988) *Periódico*, Caracas.

Grupo Feminista Miércoles (1978–91) Manuscripts and interviews.

Grupo Feminista Persona (1978–80) Manuscripts and interviews.

Kaplan, T. (1982) 'Female consciousness and collective action: The case of Barcelona, 1910–1918', *Signs* 7(3): 545–66.

Karst, K., Schwartz, M. and Schwartz, A. (1973) *The Evolution of Law in the Barrios of Caracas*, Los Angeles: Latin American Center, University of California.

Leacock, E. (1987) 'Women, power, and authority', in L. Daube, E. Leacock and S. Ardener (eds) *Visibility and Power: Essays on Women in Society and Development*, New York: Oxford University Press.

—— and Safa, H. (1986) *Women's Work*, South Hadley MA: Bergin Publishers.

Leff, E. (1991) 'El movimiento social ecologista en México', in M. P. García-Guadilla (ed.) *Ambiente, Estado y Sociedad: Crisis y Conflictos Socioambientales en América Latina y Venezuela*.

Logan, K. (1984) *Haciendo Pueblo: The Development of a Guadalajaran Suburb*, Tuscaloosa: University of Alabama Press.

—— (1988) 'Women's political activity and empowerment in Latin American urban movements', in G. Gmelch and W. P. Zenner (eds) *Urban Life*, Prospect Heights, III: Waveland.

—— (1990) 'Women's participation in urban protest', in J. Foweraker and A. Craig (eds) *Popular Movements and Political Change in Mexico*, Boulder: Lynne Reinner.

Lomnitz, L. (1975) *Como Sobreviven los Marginados*, Mexico: Siglo Veintiuno Editores.

Lope Bello, N. (1979) *La Defensa de la Ciudad*, Caracas: Universidad Simón Bolívar Instituto de Estudios Regionales y Urbanos.

Los Círculos Femeninos Populares (1990) *Boletín*, Caracas: El Pueblo.

Lozano Pardinas, D. and Padilla Tieste, C. (1988) 'La participación de la mujer en los movimientos urbano-populares', in L. Gabayet *et al. Mujeres y Sociedad*, Guadalajara: El Colegio de Jalisco.

MARNR (1991) *Políticas y Estrategias del Gobierno Nacional sobre Cuestiones Ambientales*, Ministerio del Ambiente y de los Recursos Naturales Renovables, Caracas: November.

—— (1992) *Informe de Venezuela a la Conferencia de las Naciones Unidas sobre el Medio Ambiente y el Desarrollo* (Versión Preliminar Corregida), Ministerio del Ambiente y de los Recursos Naturales Renovables.

242

Melucci, A. (1985) 'The symbolic challenge of contemporary social movements', *Social Research* 52(4).
—— (1988) 'Social movements and the democratization of everyday life', in J. Keane (ed.) *Civil Society and the State*, London: Verso.
—— (1989) *Nomads of the Present: Social Movements and Individual Needs in Contemporary Society*, London: Verso.
Merola, G. (1985) 'Feminismo, un Movimiento Social', *Revista Nueva Sociedad*, Caracas: Agosto: 112–17.
Ministerio del Estado para la Participación de la Mujer en el Desarrollo (1980) *La Mujer y los Medios de Comunicación Social en Venezuela.*
Molyneux, M. (1985) 'Mobilization without emancipation? Women's interest, state and revolution in Nicaragua', in D. Slater (ed.) *New Social Movements and the State in Latin America*, Amsterdam: CEDLA.
Naim, M. and Pioango, R. (eds) (1984) *El Caso Venezuela: Una Ilusión de Armonia*, Caracas: Ediciones IESA.
Norris, R. and Capobasso, L. (1992) *Parks and the Private Sector: Building NGO Relationship.* IV World Congress on National Parks and Protected Areas. Caracas. 10–12 February.
Ovalles, O. (1987) 'Movimientos de cuadro de vida en la Venezuela urbana actual', in L. Gomez, (ed.) *Crisis y Movimientos Sociales en Venezuela*, Caracas: Trópicos.
—— (1989) '¿Explosión social o redes de solidaridad?: un enfoque urbano del problema', *Revista Tierra Firme* Año (25).
Peattie, L. (1972) *The View from the Barrio*, Ann Arbor: The University of Michigan Press.
Pereira, I. E. Z. (1989) *La Mujer en Venezuela.* Oficina del Ministro de Estado para la Promoción de la Mujer. Edicionaes Comisión Presidencial para la Reforma del Estado, Caracas: COPRE.
Primera Jornada Binacional (1990) 'Mujer y ambiente', 'La mujer protagonista en la defensa y conservatión del ambiente', Universidad Nacional Experimental Táchira, San Cristobal: Estado Táchira, Venezuela. 1–4 November.
Puig, J. (1990) 'Larga marcha por la vida', *Revista Nuestro Ambiente.* 15(3).
Quadri, G. (1990) 'Una breve crónica del ecologismo en México', *Revista Ciencias* (4).
Rao, B. (1989) 'Struggling for production conditions for emancipation: women and water in rural Maharastra', *Capitalism, Nature, Socialism: A Journal of Socialist Ecology* 1(2): 20–33.
—— (1991) *Dominant Constructions of Women and Nature in Social Science Literature*, CES/CNS Pamphlet 2.
Ray, T. (1969) *The Politics of the Barrios of Venezuela*, Berkeley: University of California Press.
Revista ARGOS (1990) No. 11. (Número especial aniversario sucesos de febrero 1989). Universidad Simón Bolívar.
Rey, J. C. (1987) 'El futuro de la democracia en Venezuela' in *Venezuela Hacia el 2000: Desafíos y Opciones*, Editorial Nueva Sociedad, Caracas: ILDISD-UNITAR/PROFAL.
Santana, E. and Perrone, L. (1991) 'La visión ambiental desde el movimiento vecinal', in M. P. García Guadilla (ed.) *Ambiente, Estado y Sociedad: Crisis y Conflictos Socioambientales en América Latina y Venezuela.*
Saporta, N., Navarro-Aranguran, M., Chuchryk, P. and Alvarez, S. (n.d.) *Feminisms in Latin America: From Bogotá to Taxco.* Manuscript.

REFERENCES

Shiva, V. (1987) *Forestry Crisis and Forestry Myths: A Critical Review of Tropical Forests, A Call for Action*. Malaysia: World Rainforest Movement.

—— (1989) *Staying Alive: Women, Ecology and Survival in India*. New Delhi: Kali for Women.

Slater, D. (ed.) (1985) *New Social Movements and the State in Latin America*. Amsterdam: CEDLA.

Sunkel, O. (1983) 'La interacción entre los estilos de desarrollo y el medio ambiente en América Latina', in M. Botero (ed.) *Ecodesarrollo, el Pensamiento del Decenio*, Bogota: INDERENA/PNUMA.

Torres, S. and Arenas, F. (1986) 'Medio ambiente y región: ámbitos claves para la participación en la gestión democrática de un desarrollo nacional sostenible', *Revista Interamericana de Planificación* 20(77).

Uribe, G. and Lander, E. (1991) 'Acción social, efectividad simbólica y nuevos ámbitos de lo político en Venezuela', in M. P. García Guadilla (ed.) *Ambiente, Estado y Sociedad: Crisis y Conflictos Socioambientales en América Latino y Venezuela*.

USB-ILDIS (1987) *Primeras Jornadas de Evaluación de las Organizaciones Ambientalistas*, Caracas.

Vola, E. (1987) 'O movimento ecológico no Brasil (1974–1986): do ambientalismo a ecopolítica', *Revista Brasileira de Cienciais Sociais*, 3.

World Commission on Environment and Development (1987) *Our Common Future*, Oxford: Oxford University Press.

World Women's Congress for a Healthy Planet (1991) *Declaración de las Mujeres de América Latina y el Caribe*, Miami, Florida: 8–12 November.

4 'WE LEARNED TO THINK POLITICALLY'

Abreu, A. R. (1986) 'O Avêsso da Moda: Trabalhar a Domicílio da Industria nade na confecção', São Paulo: HUCITEC.

Alvarez, S. (1985) 'The politics of gender and the *Abertura* process: alternative perspectives on women and the state in Latin America', paper presented at the XII International Convention of the Latin American Studies Association, Albuquerque, New Mexico.

—— (1990) *Engendering Democracy in Brazil*, Princeton: Princeton University Press.

Alves, B. (1980) *Ideologia e Feminismo*, Petrópolis: Vozes.

Barreira, I. and Stroh, P. (1983) 'O movimento dos desempregados nas ruas: uma política de tempo e lugar?', paper presented at the VII Annual Meeting of the National Association of Post-Graduate and Research in Social Sciences, Aguas de São Pedro.

Barroso, C. (1982) *Mulher, Sociedade e Estado no Brasil*, São Paulo: UNICEF/Brasiliense.

Betto, F. (1984) 'Riqueza e pobreza no Brasil', in F. Betto (ed.) *Desemprego, Causas e Consequências*, São Paulo: Paulinas.

Boran, A. (1989) 'Popular movements in Brazil: a case study of the Movement for the Defence of Favelados in São Paulo', *Bulletin of Latin American Research*, 8(1): 83–109.

Branco, C. (1975) *Introdução à Revolução de 1964*, Rio de Janeiro: Artenova.

—— (1976) *Os militares no Poder*, Rio de Janeiro: Nova Fronteira.

Brant, V. C. (1981) 'Da resistência aos movimentos sociais: a emergência das classes populares em São Paulo', in P. Singer and V. Brant *São Paulo: O Povo em Movimento*, Petrópolis: Vozes/Cebrap: 9–27.

Brant, V. C. *et al.*, (1989) *São Paulo, Trabalhar e viver*, São Paulo: Brasiliense.

Brasileiro, A. (1982) 'Politicas sociais para areas urbanas: possibilidades', in E. Diniz (ed.) *Politicas Púbilicas para Areas Urbanas. Dilemas e Alternativas*, Rio de Janeiro: Zahar: 43–66.

Caldeira, T. (1982) 'Imagens do poder e da sociedade', Master's Dissertation, Department of Social Sciences, Universidade de São Paulo.

—— (1987) 'Mujeres, cotidianidad y politica', in E. Jelin (ed.) *Ciudadania e identidad: las mujeres en los movimientos sociales latino-américanos*, Geneva: UNRISD, 75–128.

Cardoso, F. (1979) 'On the characterization of authoritarian regimes in Latin America', in D. Collier (ed.) *The New Authoritarianism in Latin America*, Princeton: Princeton University Press: 33–57.

—— (1986) 'Entrepreneurs and the transition process: the Brazilian case' in G. O'Donnell, P. Schmitter and L. Whitehead (eds) *Transition from Authoritarian Rule*, Baltimore: The Johns Hopkins University Press: 137–53.

Cardoso, R. (1983a) 'Movimentos sociais urbanos: balanço crítico', in B. Sorj and M. de Almeida (eds) *Sociedade e Politica no Brasil Pós-64*, São Paulo: Brasiliense: 215–39.

—— (1983b) 'Novas formas de participação politica: as mulheres no Brasil', paper presented at the Technical Regional Seminar on Women and Families of the Latin American Urban Popular Strata, Santiago.

Castells, M. (1977) *The Urban Question*, London: Edward Arnold.

—— (1979) *City, Class and Power*, London: Macmillan.

—— (1983) *The City and the Grassroots*, London: Edward Arnold.

Castro, P. (1983) 'Indícios na Teia da Mobilização Popular Urbana: O caso Acarí', in R. Boschi (ed.) *Movimentos Coletivos no Brasil Urbano*, Rio de Janeiro: Zahar: 75–102.

Chiriac, J. and Padilha, S. (1982) 'Características e limites das organizaçães de base femininas', in M. Bruschini and F. Rosemberg (eds) *Trabalhadoras do Brasil*, São Paulo: Fundação Carlos Chagas/Brasiliense: 191–203.

Corcoran-Nantes, Y. (1988) 'Women in grassroots protest politics in São Paulo, Brazil', unpublished PhD Thesis, University of Liverpool.

de Camargo, C. *et al.* (1981) 'Comunidades Eclesiais de Base' in P. Singer and V. Brant (eds) *São Paulo: O Povo em Movimento*, Petrópolis: Cebrap/Vozes: 59–81.

Departamento Intersindical de Estatística e Estudos Sócio-Economicos (DIEESE) (1984) *Boletim DIEESE*, May, São Paulo.

Diniz, E. (1983) 'Favela, associativismo e participação social', in R. Boschi (ed.) *Movimentos Coletivos no Brasil Urbano*, Rio de Janeiro: Zahar: 27–74.

dos Santos, C. (1981) *Movimentos Urbanos no Rio de Janeiro*, Rio de Janeiro: Zahar.

Duarte, V. and Yasbeck, M. (1982) 'Atuaçáco de Igreja frente aos movimentos populares: uma revisão crítica da literatura', paper presented at the VI Annual Meeting of the National Association of Post-Graduate and Research in Social Science, Fribourg, October.

EMPLASA (Empresa Metropolitana de Planejamento da Grande São Paulo) (1982) *A Grande São Paulo Hoje*, São Paulo.

—— (1985) *Sumário de Dados da Grande São Paulo – 1984*, São Paulo.

Evers, T. (1982) 'Os movimentos sociais urbanos: o caso do movimento do custo de vida' in J. Moisés (ed.) *Alternativas Populares da Democracia – Brasil anos-80*, Petrópolis, Cedec/Vozes: 73–98.

Evers, T., Muller-Platenberg, C. and Spessart, S. (1982) 'Movimentos de bairro e estado: lutas na esfera da reprodução na America Latina', in

REFERENCES

J. Moisés *et al.* (eds) *Cidade, Povo e Poder*, Rio de Janeiro: Cedec/Paz e Terra.

Flynn, P. (1979) *Brazil: a Political Analysis*, Boulder: Westview Press.

Gohn, M. (1985) *A Força da Periferia*, Petrópolis: Vozes.

Grupo de Educação Popular/PUC-SP (1984) *Que história é essa? Conselhos Populares* GEP-URPLAN, No. 1, October, São Paulo: 28–43.

Jacobi, P. (1983) 'Movimentos populares urbanos e resposta do estado: auto-nomia e controle vs. cooptação e clientelismo', in R. Boschi (ed.) *Movimentos Coletivos no Brasil Urbano*, Rio de Janeiro: Zahar: 145–79.

Jacobi, J. and Nunes, E. (1981) 'Movimentos por melhores condiçães de saúde: Zona Leste de São Paulo – a secretaria de saúde e o povo', paper presented at the V Annual Meeting of the National Association of Post-Graduate and Research in Social Science, Fribourg: 20–3.

Jaquette, J. (ed.) (1989) *The Women's Movement in Latin America*, London: Unwin and Hyman.

Kowarick, L. (1979) *A Espolição Urbana*, Rio de Janeiro: Paz e Terra.

—— (1982) 'O preço do progresso: crescimento econômico, pauperização e espoliação urbana' in J. Moisés *et al.* (eds) *Cidade, Povo e Poder*, São Paulo: Cedec/Paz e Terra: 30–48.

—— (1985) 'The pathways to encounter: reflections on social struggle in São Paulo' in D. Slatter (ed.) *New Social Movements and the State in Latin America*, Amsterdam: CEDLA: Latin American Studies 29: 73–93.

Lagoa, A. (1985) *Como se Faz para Sobreviver com um Salário Mínimo*, Petrópolis: Vozes.

Langenbuch, J. (1971) *A Estruturação da Grande São Paulo*, Rio de Janeiro: Instituto Brasileiro de Geografia e Estatística (IBGE).

Lesbaupin, I. (1980) 'A igreja Católica e os movimentos populares urbanos', *Revista Religião e Sociedade*, 5: 189–98.

Lopes, J. (1980) 'Igreja e movimentos populares urbanos – comentários ao artigo de Ivo Lesbaupin', *Revista Religião e Sociedade*, 5: 199–204.

Machado, L. M. V. (1988) 'The participation of women in the Health Movement of Jardim Nordeste, in the Eastern Zone of São Paulo, Brazil: 1976–1985', *Bulletin of Latin American Research*, 7(1): 47–63.

Mainwaring, S. (1986) *The Catholic Church and Politics in Brazil – 1916–1985*, Stanford: Stanford University Press.

Moser, C. (1987) 'Women, human settlements and housing: a conceptual framework for analysis and policy making', in C. Moser and L. Peake (eds) *Women, Human Settlements and Housing*, London: Tavistock: 12–32.

Nunes, E. (1982) 'Inventário dos quebra-quebras nos trens e ônibus em São Paulo e Rio de Janeiro, 1977–1981' in J. Moisés *et al.* (eds) *Cidade, Povo e Poder*, São Paulo: Cedec/Paz e Terra: 92–108.

Rede Mulher (1985) *Retrato dos clubes de mães e grupos de mulheres da Zona Leste de São Paulo*, Pesquisa Avaliação dos clubes e grupos de mães da cidade de São Paulo, Documento No. 3.

Rolnik, R., Kowarick, L. and Somekh, N. (eds) (1989) *São Paulo, Crise e Mudança*, Municipal Department of Planning of São Paulo, São Paulo: Brasiliense.

Sabóia, J. (1985) *Salário Mínimo. A Experiência Brasileira*, Porto Alegre: LPM Editora.

Saffioti, H. (1978) *Women in Class Society*, New York: Monthly Review Press.

Sarti, C. (1989) 'The panorama of Brazilian feminism', *New Left Review*, 173, January/February: 75–90.

Schmink, M. (1981) 'Women in Brazilian Abertura politics', *Signs* 7(11): 115–34.

Singer, P. (1981) 'O feminino e o feminismo' in P. Singer and V. Brant (eds) *São Paulo: O Povo em Movimento*, Petrópolis: Vozes/Cebrap: 109–41.

Taube, M. (1986) *De Migrantes a Favelados*, Campinas: Editora da Universidade Estadual de Campinas.

Telles, V. and Bava, S. (1981) 'O movimento dos ônibus; articulação de um movimento reinvindicatório de periferia', *Espaço e Debates*, 1(1): 77–101.

Valladares, L. (1981) 'Quebra-quebras na construção civil: o caso dos operários do Metrô do Rio de Janeiro', *Dados*, 24 (1): 61–83.

Velasco e Cruz, S. and Martins, C. (1983) 'De Castello a Figueiredo: uma incursão na pré-história da Abertura' in B. Sorj and M. de Almeida (eds) *Sociedade e Política no Brasil Pós-64*, São Paulo: Brasiliense: 13–61.

Vink, N. (1985) 'Base communities and urban social movements – a case study of the metalworkers' strike in 1980, São Bernardo, Brasil' in D. Slater (ed.) *New Social Movements and the State in Latin America*, Amsterdam: CEDLA Latin American Studies 29: 95–125.

Viola, E. and Mainwaring, S. (1985) 'Brazil and Argentina in the 1980s', *Journal of International Affairs*, 38(2): 193–219.

5 WOMEN'S POLITICAL PARTICIPATION IN *COLONIAS POPULARES* IN GUADALAJARA, MEXICO

Alba, F. (1982) *The Population Trend of Mexico: Trends, Issues and Policies*, New Brunswick: Transaction Books.

Alonso, J. (1988) 'El papel de las convergencias de los movimientos sociales en los cambios del sistema político Mexicano', paper given at the Conference on Latin America held at the Centre for US–Mexican Studies, San Diego: University of California, 23–25 March.

Alvarez, S. (1984) *The Politics of Gender in Latin America: Comparative Perspectives on Women*, PhD Thesis, Yale University.

Bassols, M. and Delgado, A. (1986) 'La ONOP y las organizaciones de colonos: una aproximación a un estudio', mimeo.

Bruce, J. and Kohn, M. (eds) (1986) *Learning About Women and Urban Services in Latin America and the Caribbean*, New York: The Population Council Inc.

Carillo, T. (1990) 'Women and independent unionism in the garment industry', in J. Foweraker and A. Craig (eds) *Popular Movements and Political Change in Mexico*, Boulder and London: Lynne Reinner Publishers.

Castells, M. (1982) 'Squatter settlements in Latin America: a comparative analysis of urban social movements in Chile, Peru and Mexico', in H. I. Safa (ed.) *Towards a Political Economy of Urbanisation in Third World Countries*, Delhi: Oxford University Press.

CEPAL (1984) *La Mujer en el Sector Popular Urbano: América Latina y el Caribe*, Santiago: United Nations Publications.

Chant, S. (1992) 'Women's work and household change in Mexico in the 1980s', in N. Harvey (ed.) *The Difficult Transition: The Challenge of Reform in Mexico*, London: I. B. Tauris.

CONAMUP (1988) *III Encuentro Nacional de Mujeres de la CONAMUP*, Mexico DF: Ediciones Pueblo.

Cortina, R. (1984) 'La mujer y el magisterio en la ciudad de México' in *Fem* 36: 37–40.

Craske, N. (1992) 'Las mujeres y la CNOP: el caso de la Federación de Colonias Populares de Jalisco', in A. Massolo (ed.) *Mujeres y Ciudades: Participación Social, Trabajo y Vida Cotidiana*, Mexico DF: PIEM/El Colegio de México.

REFERENCES

Davidoff, L. (n.d.) 'Beyond the public and private: thoughts on feminist history in the 1990s', mimeo, Department of Sociology, University of Essex.

Foweraker, J. and Craig, A. (eds) (1990) *Popular Movements and Political Change in Mexico*, Boulder and London: Lynne Reinner Publishers.

González de la Rocha, M. (1988) 'Economic crisis, domestic reorganisation and women's work in Guadalajara, Mexico', *Bulletin of Latin American Research* 7(2): 207–23.

—— and Escobar Itapi, A. (1988) 'Crisis of adaptation: households of Guadalajara', *Texas Papers on Mexico*, 88.04, Austin: University of Texas at Austin.

Hellman, J. (1983) *Mexico in Crisis*, New York: Holmes and Meier Publishers.

Hess, B. and Marx Ferree, M. (eds) (1987) *Analysing Gender: A Handbook of Social Science Research*, Newbury Park: Sage Publications.

INEGI (1991) XI Censo 1990: Resultados Definitivos de Jalisco, INEGI, Aguascalientes.

Jaquette, J. (1980) 'Female political participation in Latin America', in J. Nash and H. Safa (eds) *Sex and Class in Latin America*, New York: Bergin Publishers.

—— (ed) (1989) *The Women's Movement in Latin America*, London: Unwin Hyman.

Jelin, E. (ed.) (1990) *Women and Social Change in Latin America*, London: UNRISD/Zed Books.

Jimenez, E. (1988) 'New forms of community participation in Mexico City: success or failure?', *Bulletin of Latin American Studies* 7(1): 17–31.

Kaplan, T. (1982) 'Female consciousness and collective action: the case of Barcelona 1910–1918', *Signs* 7(3).

Knight, A. (1986) *The Mexican Revolution*, Cambridge: Cambridge University Press.

—— (1990) 'Historical continuities in social movements', in J. Foweraker and A. Craig (eds) *Popular Movements and Political Change in Mexico*, Boulder and London: Lynne Reinner Publishers.

Levy, D. and Székely, G. (1987) *Mexico: Paradoxes of Stability and Change*, Boulder and London: Westview Press.

Logan, K. (1984) *Haciendo Pueblo: The Development of a Guadalajara Suburb*, Alabama: University of Alabama Press.

—— (1988) 'Latin America urban mobilization: women's participation and self-empowerment', in G. Gmelch and W. P. Zenner (eds), *Urban Life: Readings in Urban Anthropology*, Prospect Heights: Waveland Press.

—— (1989) 'Comment', in W. Cornelius, J. Gentleman and P. Smith (eds) *Mexico's Alternative Political Futures*, Monograph Series 30, San Diego: Center for US–Mexican Studies, University of California.

Lozano, D. and Padilla, C. (1988) 'La participación de la mujer en los movimientos urbano-populares', in L. Gabyet, P. García, M. González de la Rocha, S. Lailson and A. Escobar (eds) *Mujeres y Sociedad: Salario, Hogar y Acción Social en el Occidente de México*, Guadalajara: El Colegio de Jalisco/CIESAS.

Massolo, A. (1983) 'Las mujeres en los movimientos sociales urbanos de la ciudad de México', *Iztapalapa* June–Dec.

—— (1989a) 'Mujer y política urbana: la desconocida de siempre, la siempre presente', paper given at the forum *Mujer y Políticas Públicas* MAS/Fundación F. Ebert 25–8 April. Mexico DF.

—— (1989b) 'Ciudadanía y defensa de la vida: algo más de las mujeres en la lucha urbana' paper given at the round-table discussion *La Lucha Urbana en Guadalajara y México: Política, Actores y Nuevos Retos* at the International Bookfair, Guadalajara, 30 November.

—— and Díaz, L. (1984) 'La participación de las mujeres en los movimientos sociales urbanos en la ciudad de México: un proyecto de investigación', in CEPAL *La Mujer en el Sector Popular Urbano: América Latina y el Caribe*, Santiago: United Nations Publications.

—— and Schteingart, M. (eds) (1987) *Participación Social, Reconstrución y Mujer: el Sismo de 1985*, Documentos de Trabajo 1, Mexico DF: PIEM/El Colegio de México/UNICEF.

Mohr Peterson, J. (1990) 'Gender subjectivity and popular urban movements in Mexico', presented at *Fronteras, Puentes y Barreras* Conference, Monterrey, March.

Molyneux, M. (1985) 'Mobilization without emancipation? Women's interests, the state and revolution in Nicaragua', *Feminist Studies* 11(2): 227–54.

Pateman, C. (1980) 'Women and consent', in *Political Theory* 8(2): 149–68.

—— (1983) 'Feminism and democracy', in G. Duncan (ed.) *Democratic Theory and Practice*, Cambridge: Cambridge University Press.

Perelli, C. (1989) 'Putting conservatism to good use: women and unorthodox politics in Uruguay, from breakdown to transition', J. Jaquette (ed.) *The Women's Movement in Latin America*, London: Unwin Hyman.

Peterson, J. (1991) 'The redefining of citizenship by the state: the distribution of services in Guadalajara, Mexico', mimeo, Department of Sociology, University of Texas at Austin.

Portes, A. (1985) 'Latin American class structures: their composition and change during the last decade', *Latin American Research Review* 20(3): 7–39.

Prieto, A. M. (1987) 'Mexico's national coordinators in a context of economic crisis', in B. Carr and R. Anzaldúa Montoya (eds) *The Mexican Left, the Popular Movements and the Politics of Austerity*, San Diego: Center for US–Mexican Studies, University of California.

Quintana, S. V. M. (1980) *Educación Popular y Movimientos Reivindicativos Urbanos: El Caso de Sta Margarita*, Guadalajara: SEDOC.

Ramírez Saíz , J. M. (1987) *Política Urbana y Lucha Popular*, Mexico DF: UAM-Xochilmilco.

—— (1990) 'Urban struggles and their political consequences', in J. Foweraker and A. Craig *Popular Movements and Political Change in Mexico*, Boulder and London: Lynne Reinner Publishers.

Ramos Oranday, R. (1985) 'Oposición y abstecionismo en las elecciones presidenciales 1964–1982', in P. González Casanova (ed.) *Las Elecciones en México: evolución y perspectivas*, Mexico DF: Siglo XXI.

Regalado, J. (1986) 'El movimiento popular independiente de Guadalajara', in J. Tamayo (ed.) *Perspectivas de los Movimientos Sociales en la Región Centro-Occidente*, Mexico DF: Editorial Línea.

Romero, L.(1986) 'El movimiento fascista en Guadalajara' in J. Tamayo (ed.) *Perspectivas de los Movimientos Sociales en la Región Centro-Occidente*, Mexico DF: Editorial Línea.

Salinas, C. (1989) 1st State of the Nation Address reproduced in *Mexico Journal* December.

Sánchez Susarrey, J. and Medina Sánchez, I. (1987) *Jalisco desde la Revolución: Historía Política*, 9, Guadalajara: University of Guadalajara.

Stephen, L. (1989) 'Popular feminism in Mexico: women in the urban popular movement', *Zeta Magazine*, December.

Story, D. (1986) *The Mexican Ruling Party: Stability and Authority*, New York: Praeger.

Tiano, S. (1984) 'The public–private dichotomy: theoretical perspectives on "women in development"', *The Social Science Journal* 21(4): 11–28.

Uñikel, L. (1978) *El Desarrollo Urbano de México: diagnóstico e Implicaciones Futuras*, Mexico City: El Colegio de México.

Vázquez, D. (1989a) *Guadalajara: Ensayos de Interpretación*, Guadalajara: El Colegio de Jalisco.

—— (1989b) 'Rural–urban land conversion on the periphery of Guadalajara', in A. Gilbert (ed.) *Housing and Land in Urban Mexico*, Monograph Series 31, San Diego: Center for US–Mexican Studies, University of California.

6 FEMALE CONSCIOUSNESS OR FEMINIST CONSCIOUSNESS?

Blachman, M. (1973) *Eve in an Adamocracy*, Occasional Papers No. 5, New York University.

Blay, E. A. (1980) 'Mulheres e movimentos sociais urbanos no Brasil: anistia, custo de vida e creches', *Encontros com a Civilização Brasileira*, 26: 63–70.

Bourque, S. C. (1985) 'Urban activists: paths to political consciousness in Peru', in S. C. Bourque and D. Robinson Divine (eds) *Women Living Change*, Philadelphia: Temple University Press: 25–56.

—— and Grossholtz, I. (1986) 'Politics as unnatural practice: political science looks at female participation', in J. Siltanen and M. Stanworth (eds) *Women in the Public Sphere: a Critique of Sociology and Politics*, London: Hutchinson.

Brydon, L. and Chant, S. (1989) *Women in the Third World*, Aldershot: Edward Elgar.

Cardoso, R. C. L. (1984) 'Movimentos sociais urbanos: balanço critico', in B. Sorj and M. H. de Almeida, (eds) *Sociedade e Politica no Brasil Pos-1964*, São Paulo: Editora Brasiliense: 226–39.

Carroll, B. A. (1989) 'Women take action!: women's direct action and social change', *Women's Studies International Forum*, 12: 3–24.

Censo Demographico do Brasil 1990.

Chaney, E. (1979) *Supermadre: Women in Politics in Latin America*, Austin: University of Texas Press.

Chuchryk, P. (1989) 'Subversive mothers: the opposition to the military regime in Chile', in S. M. Charlton, J. Everett and K. Staudt, (eds) *Women, the State and Development*, Albany: State University of New York Press.

Clements, B. E. (1982) 'Working class and peasant women in the Russian revolution', *Signs*, 8(2).

Corcoran-Nantes, Y. (1988) 'Women in grass roots protest politics in São Paulo, Brazil', Unpublished PhD Thesis, Department of Sociology, University of Liverpool.

—— (1990) 'Women and popular urban social movements in São Paulo, Brazil', *Bulletin of Latin American Research*, 9(2): 249–64.

Cutrufelli, M. R. (1983) *Women of Africa: Roots of Oppression*, London: Zed Books.

El Saadawi, N. (1985) 'Politics: Great Britain', in *Women: A World Report*, London: Methuen: 241–70.

Evers, T., Muller-Plantenberg, C. and Spessart, S. (1982) 'Movimentos de bairro e estado: lutas na esfera de reprodução na America Latina', in J. A. Moises (ed.) *Cidade, Povo e Poder*, Rio de Janeiro: Paz e Terra: 110–60.

Filet Abreu de Souza, J. (1980) 'Paid domestic service in Brazil' in *Latin American Perspectives*, 24(7).

Jaquette, J. (1980) 'Female political participation in Latin America', in J. Nash and H. Safa (eds) *Sex and Class in Latin America*, New York: Bergin Publishers.

—— (ed.) (1989) *The Women's Movement in Latin America: Feminism and the Transition to Democracy*, London: Unwin Hyman.

Kaplan, T. (1982) 'Female consciousness and collective action: the case of Barcelona, 1910–1918', *Signs*, 7(3): 545–66.

Kucinski, B. (1982) *Abertura uma historia de uma crise*, São Paulo: Brasil Debates.

Machado, L. (1988) 'The participation of women in the health movement of

Jardim Nordeste, in the Eastern Zone of São Paulo, Brazil: 1976–1985', *Bulletin of Latin American Research* 7(1): 47–63.

Mattelart, M. (1980) 'The feminine version of the coup d'etat' in J. Nash and H. I. Safa (eds) *Sex and Class in Latin America*, New York: Bergin Publishers.

Mies, M. (1988) *Women: The Last Colony*, London: Zed Books.

Molyneux, M. (1985) 'Mobilization without emancipation? Women's interest, state and revolution in Nicaragua', in D. Slater (ed.) *New Social Movements and the State in Latin America*, CEDLA Amsterdam: Latin American Studies 29: 233–60.

Moser, C. O. N. (1987) 'Mobilisation is women's work: the struggle for infrastructure in Guayaquil, Ecuador' in C. O. N. Moser and L. Peake, (eds) *Women, Human Settlements and Housing*, London and New York: Tavistock Publications: 166–94.

Offe, K. (1985) 'New social movements: challenging the boundaries of institutional politics', *Social Research*, 52: 817–68.

Randall, V. (1987) *Women and Politics*, Basingstoke: Macmillan.

Safa, H. I. (1990) 'Women's social movements in Latin America', *Gender and Society*, 4(3).

Siltanen, J. and Stanworth, M. (1984) 'The politics of private woman and public man' in J. Siltanen and M. Stanworth (eds) *Women in the Public Sphere: a Critique of Sociology and Politics*, London: Hutchinson: 183–208.

Slater, D. (ed.) (1985) *New Social Movements and the State in Latin America*, Amsterdam: CEDLA: Latin American Studies 29.

Stevens, E. (1973) 'Machismo and marianismo', *Society*, 10(6).

Tabak, F. (1983) *Autoritarismo e Participação Politica da Mulher*, Rio de Janeiro: Edições Graal Ltda.

Thomis, M. I. and Grimmett, J. (1982) *Women in Protest*, London: Croom Helm.

Tilly, L. A. (1981) 'Paths of proletarianization: organization of production, sexual division of labour and women's collective action', *Signs*, 7(2): 400–17.

Wanderley, L. E. (1980) 'Movimientos sociales populares: aspectos económicos e políticos', *Encontros com a Civilização Brasileira*, 25: 107–31.

7 TOUCHING THE AIR

Allende, I. (1982) *La Casa de los Espíritus*, Barcelona: Plaza y Janés; trans. by Mazda Bogin, 1985, Jonathan Cape.

Agosín, M. (1987) *Scraps of Life, Chilean arpilleras*, London: Zed Books.

Angwell, A. (1990) 'The Chilean elections of 1989 and the politics of the transition to democracy', *Bulletin of Latin American Studies*, 9(1): 1–23.

Berenguer, C., et al., (compilers) (1990) *Escribir en los bordes. Congreso Internacional de Literatura Femenina Latinoamericana 1987*, Santiago: Editorial Cuarto Propio.

Boyle, C. M. (1986) 'Images of women in contemporary Chilean theatre', *Bulletin of Latin American Research*, 5(2): 81–96.

Chaney, E. (1974) 'The mobilization of women in Allende's Chile', in J. Jaquette (ed.) *Women in Politics*, New York: John Wiley & Sons: 267–80.

Chuchryk, P. M. (1989) 'Feminist anti-authoritarian politics: the role of women's organizations in the Chilean transition to democracy', in J. Jaquette (ed.) *The Women's Movement in Latin America. Feminism and the Transition to Democracy*, Boston: Unwin Hyman.

Five Studies on the Situation of Women in Latin America (1983), Santiago: CEPAL.

Gálvez, T. and Todaro, R. (1990) 'Chile: women and the unions', in E. Jelin (ed.) *Women and Social Change in Latin America*, London and New Jersey: Zed Books: 115–34.

Gissi Bustos, J. (1980) 'Mythology about women, with special reference to Chile', in J. Nash and H. I. Safa, (eds) *Sex and Class in Latin America*, New York: Bergin Publishers: 30–45.

Kirkwood, J. (1990) *Ser Política en Chile. Los Nudos de la Sabiduría Feminista*, Santiago: Editorial Cuarto Propio.

—— and Molina, N. (1986) *Lo Femenino y lo Democrático en el Chile de Hoy*, Santiago: Vector (Centro de Estudios Económicos y Sociales).

La Duke, B. (1985) *Compañeras. Women, Art and Social Change in Latin America*, San Francisco: City Lights Books.

Letelier, L. (n.d.) *Collage Mujeres*, Santiago: CENECA, Serie 'Identidad cultural y sociedad', Documento de Trabajo, 86.

Mattelart, M. (1976) 'Chile: the feminine version of the coup d'état', in J. Nash and H. I. Safa (eds) *Sex and Class in Latin America*, New York: Bergin Publishers: 279–301.

Meza, M. A. (n.d.) *La otra mitad de Chile*, Santiago: Centro de Estudios Sociales, Ediciones Chile y América.

Mistral, G. (1928) 'El voto femenino', *El Mercurio*, 2 June.

—— (1938) *Tala*, Buenos Aires: Editorial Sur.

Molina, N. (1990) 'El estado y las mujeres: una relación difícil', in *Transiciones. Mujeres en los Procesos Democráticos*, Santiago: Isis Internacional. Ediciones de las mujeres 13: 85–97.

Montecino, S. (1991) *Madres y Huachos. Alegorías del Mestizaje Chileno (Ensayo)*, Santiago: Editorial Cuarto Propio.

Moreno Aliste, C. (1984) *La Artesanía Urbano Marginal*, Santiago: CENECA, Documento de Trabajo, 51.

Mujeres por el Si (1988) 'Inseción. Mujer Chilena', *El Mercurio*, 4 September.

Mundo de mujer. Continuidad y cambio (1988), Santiago: Ediciones Centro de Estudios de la mujer.

Nash, J. and Safa, H. I. (eds) (1980) *Sex and Class in Latin America. Women's Perspectives on Politics, Economics and the Family in the Third World*, New York: Bergin Publishers.

Ojeda, A. (1988) *Mi Rebeldía es Vivir*, Santiago: Ediciones Literatura Alternativa.

Orellano, C. (n.d.) in M. A. Meza (ed.) *La Otra Mitad de Chile*, Santiago: Centro de Estudios Sociales, Ediciones Chile y América.

'Radio de Mujeres. Casa de la Mujer La Morada', 1991, document, Santiago.

Rowe, W. and Schelling, V. (1991) *Memory and Modernity. Popular Culture in Latin America*. London: Verso.

Santa Cruz, A. and Erazo, V. (eds) (1988), *El Cuento Feminista Latinoamericano*, Santiago: ILET.

Serrano, B. (ed.) (1988) *Poesía Prisionera. Escritura de Cinco Mujeres Encarceladas*, Santiago: Ediciones Literatura Alternativa.

Serrano, C. (1990) 'Entre la autonomía y la integración', in *Transiciones. Mujeres en los Procesos Democráticos*, Santiago: Isis Internacional. Ediciones de las Mujeres 13: 99–105.

Stranger, I. (1991) 'Cariño Malo', *Apuntes* 101: 17–24.

Taller de Investigación Teatral (1978) 'Los Payasos de la Esperanza', *Apuntes* 84: 27–80.

—— and David Benavente (1979) *Tres Marías y una Rosa* in *Teatro Chileno de la Crisis Institucional: 1973–1980 (Antología Crítica)*, Santiago: CENECA and University of Minnesota Latin American Series: 196–248.

Valenzuela, A. (1978) *The Breakdown of Democratic Regimes: Chile*, Baltimore: Johns Hopkins Press.

Vodanovic, S. (1970) 'Nos tomamos la universidad', in *Teatro. Deja que los perros ladren y Nos tomamos la universidad.*

Wolff, E. (1971) *Flores de Papel* in *Egon Wolff: Teatro – Niñamadre, Flores de Papel, Kindergarten*, Santiago: Editorial Nascimento. Trans. by Margaret Sayers 1971 Columbia: University of Missouri Press.

Zabaleta, M. (1986) 'Research on Latin American women: in search of our political independence', *Bulletin of Latin American Research*, 5(2): 97–103.

8 ADJUSTMENT FROM BELOW

Barrig, M. and Fort, A. (1987) 'La ciudad de las mujeres: pobladores y servicios. El caso de El Augustino', *Women, Low-Income Households and Urban Services Working Papers*, Lima.

BCE (Banco Central de Ecuador) (1986) Cuentas Nacionales no 8, Quito: Banco Central de Ecuador.

Commonwealth Secretariat (1990) *Engendering Adjustment for the 1990s*, London: Commonwealth Secretariat.

Cornea, G., Jolly, R. and Stewart, F. (1987) *Adjustment with a Human Face: Vol. 1*, Oxford: Oxford University Press.

—— (1988) *Adjustment with a Human Face: Vol. 2*, Oxford: Oxford University Press.

Demery, L. and Addison, T. (1987) *The Alleviation of Poverty under Structural Adjustment*, Washington: The World Bank.

Elson, D. (1987) 'The impact of structural adjustment on women: concepts and issues', paper presented at the Institute of African Alternatives Conference, City University, London.

—— (1991) 'Male bias in macroeconomics: the case of structural adjustment', in D. Elson (ed.) *Male Bias in the Development Process*, Manchester: Manchester University Press.

Evans, A. (1989) 'Gender issues in rural development economics', *Institute of Development Studies*, Discussion Paper 254.

Freire, W. (1985) 'La situación nutricional en Ecuador', *Revista Ecuador Debate*, 9, Quito: CAAP.

—— (1988) *Diagnóstico de la Situación Nutricional de la Población Ecuatoriana Menor de Cinco Anos, en 1986, Resultados Preliminares*. Quito: CONADE.

Jolly, R. (1987) 'Women's needs and adjustment policies in developing countries', an address to the Women's Development Group of the OECD, Paris.

Moser, C. O. N. (1981) 'Surviving in the suburbios', *Bulletin of the Institute of Development Studies* 12(3): 19–29.

—— (1982) 'A home of one's own: squatter housing strategy in Guayaquil, Ecuador', in A. Gilbert (ed.) *Urbanization in Contemporary Latin America*, London: John Wiley.

—— (1986) 'Women's needs in the urban system: training strategies in gender aware planning', in J. Bruce, M. Kohn and M. Schmink (eds) *Learning About Women and Urban Services in Latin America and the Caribbean*, New York: Population Council.

—— (1987) 'Mobilization is women's work: struggles for infrastructure in Guayaquil, Ecuador', in C. Moser and L. Peake (eds) *Women, Human Settlements and Housing*, London: Tavistock.

—— (1989) 'Gender planning in the Third World: meeting practical and strategic gender needs', *World Development*, 17(11): 1799–1825.

—— and Sollis, P. (1989) *The UNICEF/MSP Primary Health Care Programme*,

REFERENCES

Guayaquil, Ecuador: Evaluation of the Cisne Dos Project from a Community Perspective. Report prepared for UNICEF, New York, mimeo.
—— (1991) 'Did the project fail? A community perspective on a participatory health care project in Ecuador', *Development Practice*, 1(1): 19–33.
Selowsky, M. (1987) 'Adjustment in the 1980s: an overview of issues', *Finance and Development*, 24(2): 11–14.
Sollis, P. and Moser, C. (1990) 'A methodological framework for analysing the social costs of adjustment at the micro level: the case of Guayaquil, Ecuador, *Bulletin of the Institute of Development Studies*, 22(1): 23–30.
Suarez, J. y col (1987) *La Situación de la Salud en el Ecuador 1969–85*, Quito: MPS-INIMMS-OPS-OMS.
UNDP (1987) Regional Programme for Africa: Fourth Cycle. *Assessment of Social Dimensions of Structural Adjustment in Sub-Saharan Africa*. Paper no. RAF/86/037/A/01/42. New York: UNDP.
UNICEF (n.d.) *The Invisible Adjustment: Poor Women and the Economic Crisis*, UNICEF Americas and Caribbean Regional Office, Bogotá, Colombia.
—— (1987) *The State of the World's Children*, New York: UNICEF.
—— (1988) *La Crisis: Efectos en Niños y Mujeres Ecuatorianos*, Quito: UNICEF.
—— (1989) *The Invisible Adjustment: Poor Women and the Economic Crisis (second revised edition)*, UNICEF Americas and the Caribbean Programme: Bogotá, Colombia.
World Bank (1984) *Ecuador: An Agenda for Recovery and Sustained Growth*, Washington: The World Bank.
—— (1988) *Ecuador: Country Economic Memorandum*, Washington: The World Bank.

9 'PEOPLE HAVE TO RISE UP – LIKE THE GREAT WOMEN FIGHTERS'

Afshar, H. (ed.) (1987) *Women, State and Ideology: Studies from Africa and Asia*, London: Macmillan.
Alvarez, S. (1990) *Engendering Democracy in Brazil*, Princeton: Princeton University Press.
Andreas, C. (1985) *When Women Rebel: the Rise of Popular Feminism in Peru*, Westport: Lawrence Hill and Company.
Anthias, F. and Yuval-Davis, N. (1983) 'Contextualizing feminism: gender, ethnic and class divisions', *Feminist Review* 15: 62–75.
—— (1989) 'Introduction', in N. Yuval-Davis and F. Anthias (eds) *Woman – Nation–State*, London: Macmillan.
Barrig, M. (1986) 'Democracia emergente y movimiento de mujeres', in E. Ballón *et al.* (eds) *Movimientos Sociales y Democracia: la Fundación de un Nuevo Orden*, Lima: DESCO.
Barrios de Chungara, D. (1978) *Let me Speak! Testimony of Domitila, a Woman of the Bolivian Tin-mines*, New York: Monthly Review Press.
Blondet, C. (1987) 'Muchas vidas construyendo una identidad: las mujeres pobladoras de un barrio limeño', in E. Jelin (ed.) *Ciudadanía e Identidad: las Mujeres en Los Movimientos Sociales Latino-Americanos*, Geneva: UNRISD.
Bourque, S. (1989) 'Gender and the state: perspectives from Latin America', in S. Charlton, J. Everett and K. Staudt (eds) *Women, the State and Development*, Albany: State University of New York.
—— and Warren, K. (1981) *Women of the Andes: Patriarchy and Social Change in Two Peruvian Towns*, Ann Arbor: University of Michigan Press.

254

Bronstein, A. (1982) *The Triple Struggle: Latin American Peasant Women*, London: WOW Campaigns Ltd.

Calderón, F. and Dandler, J. (eds) (1986) *Bolivia: la Fuerze del Campesinado*, Geneva: UNRISD.

Cavassa, M. and Portugal, J. (1985) *La Mujer en Cifras*, Lima: CENDIPP.

CEAS (Comisión Episcopal de Acción Social) (1985) *Encuentro Nacional de Campesinos Cristianos*, Lima: CEAS.

Chaney, E. (1979) *Supermadre: Women in Politics in Latin America*, Austin: University of Texas Press.

Charlton, S., Everett, J. and Staudt, K. (eds) (1989) *Women, the State and Development*, Albany: State University of New York.

Cocks, J. (1989) *The Oppositional Imagination: Feminism, Critique and Political Theory*, London: Routledge.

Comisión de la Fuerza Armada (1977) *Plan de Gobierno 'Tupac Amaru' Periodo 1977–1980*, Lima: Lima S.A.

Connell, R. W. (1987) *Gender and Power*, Cambridge: Polity Press.

Deere, C. D. (1977) 'The agricultural division of labour by sex: myths, facts and contradictions in the Northern Peruvian Sierra', paper given at Joint Meeting of Latin American Studies Association/ASA, Houston, November.

—— (1985) 'Rural women and state policy: the Latin American agrarian reform experience', *World Development* 13(9): 1037–53.

—— and León, M. (1982) *Women in Andean Agriculture*: Geneva: ILO.

de Lauretis, T. (1987) *Technologies of Gender: Essays on Theory, Film and Fiction*, London: Macmillan.

Dibos Cauri, B. (1976) *Experiencia de Investigación y Promoción con el Comité Femenino de la c.c. San Pedro de Cajas, Tarma*, Lima: CONAMUP.

Eguren, F. (1988) 'Democracia y sociedad rural', in L. Pásara and J. Parodi (eds) *Democracia, Sociedad y Gobierno en el Perú*, Lima: CEDYS.

FDCP (Federación Departamental de Campesinos de Puno) (n.d.) *1⁰ Congreso Departamental de Mujeres Campesinas de Puno. 18 a 20 Marzo de 1985. Chucuito Puno: Conclusiones de Comisiones*, Puno: FDCP.

Foucault, M. (1980) *Power/Knowledge: Selected Interviews and Other Writings*, New York: Pantheon.

Franzway, S., Court, D. and Connell, R. W. (1989) *Staking a Claim: Feminism, Bureaucracy and the State*, Cambridge: Polity Press.

Fraser, N. (1989) *Unruly Practices: Power, Discourse and Gender in Contemporary Social Theory*, Cambridge: Polity Press.

García-Sayan, D. (1982) *Tomas de Tierras en el Perú*, Lima: DESCO.

Gonzales, E. (1984) *Economía de la Comunidad Campesina*, Lima: Instituto de Estudios Peruanos.

Graham, C. (1991) 'The APRA government and the urban poor: the PAIT programme in Lima's *pueblos jóvenes*', *Journal of Latin American Studies* 23(1): 91–130.

Harvey, P. (1987) 'Language and the power of history. The discourse of bilinguals in Ocongate (Southern Peru)', unpublished PhD thesis, London School of Economics, University of London.

Hernández, Z. (1986) *El Coraje de las Mineras*, Lima: Asociación Aurora Vivar.

Holloway, W. (1984) 'Gender difference and the production of subjectivity', in J. Henriques *et al. Changing the Subject: Psychology, Social Regulation and Subjectivity*, London: Methuen.

Instituto de Estudios Aymaras (1985) *Guias de Reuniones con Mujeres*, Instituto de Estudios Aymaras, Puno: Chucuito.

REFERENCES

Jackson, P. (1989) *Maps of Meaning*, London: Unwin Hyman.
—— (1991) 'The cultural politics of masculinity: towards a social geography', *Transactions of the Institute of British Geographers* New Series 16(2): 199–213.
Jaquette, J. (ed.) (1989) *The Women's Movement in Latin America: Feminism and the Transition to Democracy*, Boston: Unwin Hyman.
Jelin, E. (ed.) (1987) *Ciudadanía e Identidad: las Mujeres en los Movimientos Sociales Latino-Americanos*, Geneva: UNRISD.
Laclau, E. and Mouffe, C. (1985) *Hegemony and Socialist Strategy*, London: Verso.
Lowenthal, A. (1975) *The Peruvian Experiment*, Princeton: Princeton University Press.
MacKinnon, C. (1982) 'Feminism, Marxism and the State: an Agenda for Theory', *Signs* 7(3): 515–44.
McClintock, C. and Lowenthal, A. (eds) (1983) *The Peruvian Experiment Reconsidered*, Princeton: Princeton University Press.
Monge, C. (1989) 'La agremiación en el campo peruano: la historia de la Confederación Campesina del Perú', paper given at the XV Congress of Latin American Studies Association, Puerto Rico, September.
Moore, H. (1988) *Feminism and Anthropology*, Cambridge: Polity Press.
Mouffe, C. (1979) *Gramsci and Marxist Theory*, London: Routledge and Kegan Paul.
Mujer y Sociedad (1987) 'Hacia la autonomía total' Año 7(13): 5–6.
Muñoz, B. (1986) 'La participación de la mujer campesina en Bolivia: un estudio del Altiplano', in F. Calderón and J. Dandler (eds) *Bolivia: la Fuerza del Campesinado*, Geneva: UNRISD.
Paredes, P. and Tello, G. (1988) *Pobreza Urbana y Trabajo Femenino*, Lima: ADEC-ATC.
Pateman, C. (1988) *The Sexual Contract*, Cambridge: Polity Press.
Political Geography Quarterly (1990) Special issue on 'Gender and political geography', 9(4) October.
Portocarrero, G. (1988) 'Nacionalismo peruano: entre la crisis y la posibilidad', *Márgenes* 3: 13–45.
Portocarrero, P. (ed.) (1990) *Mujer en el Desarrollo: Balance y Perspectivas*, Lima: Centro de la Mujer Flora Tristan.
Prieto, J. (1980) *Mujer, Poder y Desarrollo en el Perú*, Lima: Edición Dorhca.
PUM Comisión Femenina (Partido Unificado Mariateguista) (1985) *Primer Congreso de Mujeres Campesinas del Departamento de Puno: Vida, Roles de la Mujer de Puno*, Lima: PUM Comisión Femenina.
Radcliffe, S. (1988) *Así es una Mujer del Pueblo: Low-income Women's Organizations under Apra, 1985–1987*, Working Paper 43, Cambridge: Centre of Latin American Studies, University of Cambridge.
—— (1990a) 'Multiple identities and negotiation over gender: female peasant union leaders in Peru', *Bulletin of Latin American Research* 9(2): 229–47.
—— (1990b) 'Ethnicity, patriarchy and incorporation into the nation: female migrants as domestic servants in Peru', *Environment and Planning D: Society and Space* 8(4): 379–93.
—— (forthcoming) *Confederations of Gender: the Mobilization of Peasant Women in Peru*, Ann Arbor: University of Michigan Press.
Rivera Cusicanqui, S. (1987) *'Oppressed but not defeated'. Peasant struggles among the Aymara and Qhechwa in Bolivia, 1900–1980*, Geneva: UNRISD.
Rowe, W. and Schelling, V. (1991) *Memory and Modernity: Popular Culture in Latin America*, London: Verso.
Ruiz Bravo, P. (1987) 'Programas de promoción y organizaciónes de mujeres',

in A. Grandón, B. Valdivia, C. Guerrero and P. Ruiz Bravo *Crísis y Organizaciones Populares de Mujeres*, Lima: Universidad Católica.

Schirmer, J. (1989) '"Those who die for life cannot be called dead": Women and human rights protest in Latin America', *Feminist Review* 32: 3–29.

Scott, A. (1990) 'Patterns of patriarchy in the Peruvian working class', in S. Stichter and J. Parpart (eds) *Women, Employment and the Family in the International Division of Labour*, London: Macmillan.

Scott, J. (1986) 'Gender: a useful category of historical analysis', *American Historical Review* 91: 1053–75.

Situación Latinoamericana (1991) 'Perú', *Situación Latinoamericana* 1, 3: 150–79.

Slater, D. (1989) *Territory and State Power in Latin America: the Peruvian Case*, London: Macmillan.

Sostres, M. and Ardaya, G. (1984) *Prácticas de Resistencia y Revindicación de la Mujer Campesina: el Caso de las 'Bartolinas'*, Montevideo.

Stephen, L. (1989) 'Not just one of the boys: from female to feminist in popular rural movements in Mexico', Paper given at the XV Congress of Latin American Studies Association, Puerto Rico, September.

Vargas, V. (1986) *Vota por tí Mujer! Reflexiones en torno a una Campana Electoral Femenista*, Lima: Centro de la Mujer Flora Tristan.

Westwood, S. (1991) 'Red star over Leicester', in M. Anwar and P. Werbner (eds) *Black and Ethnic Minority Politics in Britain*, London: Routledge.

—— (1990b) 'Racism, black masculinity and the politics of space', in J. Hearn and D. Morgan (eds) *Men, Masculinity and Social Theory*, London: Unwin Hyman.

WGSG (Women and Geography Study Group) (1984) *Geography and Gender: An Introduction to Feminist Geography*, London: Hutchinson.

INDEX